Soils Laboratory Manual

Soils Laboratory Manual

K-State Edition, Version 2.0

COLBY J. MOORBERG AND DAVID A. CROUSE

MAURICE G. COOK; KIM HAY; LARRY D. KING; H. JOSEPH KLEISS; CONNIE MCDOWELL; MICHEL D. RANSOM; NORA RANSOM; LLOYD STONE; AND JAMES A. THOMPSON

NEW PRAIRIE PRESS
MANHATTAN

Soils Laboratory Manual: K-State Edition, Version 2.0

New Prairie Press, Kansas State University Libraries, Manhattan, Kansas

Electronic edition available at: https://newprairiepress.org/ebooks/39/
Updated February 2021

Instructor copies and other supporting materials are available at: open.soilscience.info

The original version was also published as a supplement to 'An Open-Source Laboratory Manual for Introductory Undergraduate Soil Science Courses' in Natural Sciences Education, Vol. 46:170013.

Cover design by: Colby J. Moorberg with photos by John Kelley, Stan Buol, and Kansas State Research and Extension.

Contribution no. 21-205-B of the Kansas Agricultural Experiment Station.

This book was produced with Pressbooks (https://pressbooks.com) and rendered with Prince.

"We know more about the movement of celestial bodies than about the soil underfoot."

-Leonardo Da Vinci-

Contents

On the Cover

The cover image is a soil profile collage created by Colby J. Moorberg. The series name (if available), USDA *Soil Taxonomy* classification (United States Department of Agriculture Natural Resources Conservation Service, 2017), and source information for each image is below. Many of the images are available on SoilScience.info. You can view photos like these, and many others at www.flickr.com/photos/soilscience/.

Top row, left to right:

Coxville series; fine, kaolinitic, thermic Typic Paleaquults; courtesy of John Kelley

Colfax series; fine-loamy, mixed, subactive, thermic Aquic Fragiudults; courtesy of John Kelley

Dellwood series; sandy-skeletal, mixed, mesic Oxyaquic Humudepts; courtesy of John Kelley

Vaucluse series; fine-loamy, kaolinitic, thermic Fragic Kanhapludults; courtesy of John Kelley

Helena series; fine, mixed, semiactive, thermic Aquic Hapludults; courtesy of John Kelley

Leon series; sandy, siliceous, thermic Aeric Alaquods; courtesy of John Kelley

Unknown series; loamy, kaolinitic, thermic Grossarenic Kanhapludult; courtesy of Stan Buol

Second row, left to right:

Clifford series; fine, kaolinitic, mesic Typic Kanhapludults; courtesy of John Kelley

Harney series; fine, smectitic, mesic Typic Argiustolls; courtesy of Kansas State Research and Extension

Dothan series; fine-loamy, kaolinitic, thermic Plinthic Kandiudults; courtesy of John Kelley

Appling series; fine, kaolinitic, thermic Typic Kanhapludults; courtesy of John Kelley

Lynn Haven series; sandy, siliceous, thermic Typic Alaquods; courtesy of John Kelley

Cecil series; fine, kaolinitic, thermic Typic Kanhapludults; courtesy of John Kelley

Unknown series; fine, smectitic, hyperthermic Oxyaquic Vertic Hapludalf, courtesy of Stan Buol

Third row, left to right:

Altavista series; fine-loamy, mixed, semiactive, thermic Aquic Hapludults; courtesy of John Kelley

Duplin series; fine, kaolinitic, thermic Aquic Paleudults; courtesy of John Kelley

Unknown series; fine-loamy, mixed, superactive, thermic Typic Calciargid; courtesy of Stan Buol

Tonka series; fine, smectitic, frigid Argiaquic Argialbolls; courtesy of John Kelley

Biltmore series; mixed, mesic Typic Udipsamments; courtesy of John Kelley

Goldston series; loamy-skeletal, siliceous, semiactive, thermic, shallow Typic Dystrudepts; courtesy of John Kelley

Cowarts series; fine-loamy, kaolinitic, thermic Typic Kanhapludults; courtesy of John Kelley

Fourth row, left to right:

Murville series; sandy, siliceous, thermic Umbric Endoaquods courtesy of John Kelley

Tuxekan series; coarse-loamy over sandy or sandy-skeletal, mixed, superactive Typic Humicryods; courtesy of John Kelley

Westfield series; fine, kaolinitic, mesic Typic Kanhapludults; courtesy of John Kelley

Bonneau series; loamy, siliceous, subactive, thermic Arenic Paleudults; courtesy of John Kelley

Myatt series, fine-loamy, siliceous, active, thermic Typic Endoaquults; courtesy of John Kelley

Hard Labor series; fine, kaolinitic, thermic Oxyaquic Kanhapludults; courtesy of John Kelley

Rains series; fine-loamy, siliceous, semiactive, thermic Typic Paleaquults; courtesy of John Kelley

Preface

This laboratory manual was designed for students in undergraduate, introductory soil science courses, and highlights the many aspects of soil science, including: soil genesis and classification, soil physical properties, soil-water interaction, soil biology, soil chemistry, and soil fertility. The work required for the laboratory is contained within this manual with the exception of homework (quizzes, problem sets, and lab summaries). The manual includes 15 different laboratories. For any particular lab session, the schedule is as follows:

1. Review the procedures for the lab and complete any assignments.
2. Read the textbook assignment and complete the pre-lab assignment questions.
3. Conduct the lab activities and tabulate data as required.
4. Complete post-lab online quiz, problem sets, or write the lab summary as required.

The format of each laboratory is explained below.

Learning Objectives

Learning objectives are provided at the beginning of each lab that outlines the target outcomes for the lab activities and assignment. A materials list is also provided, and includes all materials required to perform the in-lab activities.

Recommended Reading and Viewing

Each lab includes a list of references that are free and available online. These resources will provide context for each lab, and will help answer most of the pre-lab assignment questions.

Pre-Lab Assignment

For each lab, you must read an applicable section in the textbook. When appropriate, you will complete a pre-lab online quiz.

Introduction

Each lab begins with a brief review of information for the lab topic.

In-Lab Activities

Some introductory material is presented, and then procedures for various activities are given. A shovel icon, is used to indicate when answers to questions in the activity are required.

Post-Lab Assignment

Each lab exercise will have one graded activity, which may include an online lab quiz, a problem set, or a lab summary. Due dates for each assignment will be designated at the time, but generally, assignments are due within one week of the day assigned.

Online Lab Quizzes

In some labs, the results of activities do not lend themselves to formal lab reports or problem sets. In labs of this type students must complete a post lab quiz via K-State Online. Quizzes are worth 20 points.

Problem Sets

Several of the topics in the course require applied math and chemistry skills. To develop and practice these skills, problem sets will be used either in lab or given as a take-home assignment. Problem sets are worth 30 points.

Lab Summaries and Reports

In several of the labs, you will explore your experimental laboratory results in more detail. Instructions for these lab summaries will be provided later. Lab summaries are worth 40 points.

About the Authors

Colby J. Moorberg, PhD, CPSS, PWS

Dr. Moorberg is an Associate Professor of Soil Science in the Kansas State University Department of Agronomy in a teaching and research appointment. He teaches AGRON 305 – Soils and AGRON 635 – Soil and Water Conservation, among other courses. He is also a coach of the K-State Soil Judging Team. He leads the K-State Hydropedology and Root Ecology Laboratory, which conducts research focused on root dynamics, rhizosphere processes, root imaging, wetland soils, and water quality. His teaching scholarly activities are focused on the development of open educational resources (OER) and soil science education. In addition to *Soils Laboratory Manual, K-State Edition* he lead the development of the open textbook, *Soil and Water Conservation – An Annotated Bibliography*. Dr. Moorberg is also a Certified Professional Soil Scientist (CPSS) and Professional Wetland Scientist (PWS).

Twitter: @ColbyDigsSoil

LinkedIn: Colby Moorberg

K-State: Colby Moorberg

David A. Crouse, PhD

Dr. Crouse is an Alumni and Distinguished Undergraduate Professor, and the Undergraduate Program Director in the Department of Crop and Soil Sciences at NC State University. Dr. Crouse's undergraduate teaching program consistently delivers a high quality learning experience to a large number of students from a diverse audience. A typical semester for his SSC – 200 Soil Science lecture and SSC – 201 Soil Science Laboratory will reach 400 students representing 60 different curricula. While his classes are high enrollment, individualized attention to the student remains a primary focus.

Twitter: @DavidCrouse

LinkedIn: David Crouse

NC State: David Crouse

Acknowledgements

The following individuals made contributions to the *Soils Laboratory Manual – K-State Edition*; its predecessor, the *Soils Laboratory Manual – NC State Edition*; or they made contributions towards the publication of the article in *Natural Sciences Education*.

Primary Authors

Colby J. Moorberg

David A. Crouse

Secondary Authors

Maurice G. Cook

Larry D. King

Joseph Kleiss

Lloyd Stone

James A. Thompson

Contributors

Kim Hay

Connie McDowell

Nora Ransom

Michel Ransom

Minor Contributors

Stephen M. Bell

Tiffany Carter

Madhubhashini "Buddhika" Galkaduwa

Kasuni Hiranthika Haputhanthri Gamage

Garrison Gundy

Che-Jen "Jerry" Hsiao

Ashley Lorence

Dorothy Menefee

Andrew Newsum

Marcie Oelke

Kyle Stropes

Luke Struckman

Erica Stuhlsatz

Steel Wagstaff

Jay Weeks

Thomas York

For Instructors

Textbook Adoption and The Soils Laboratory Manual Library

Thank you for your interest in using the *Soils Laboratory Manual* for your soil science course. If you are interested in ancillary materials developed for use with the *Soils Laboratory Manual*, then please apply for access to the *Soils Laboratory Manual* Library. The library includes many resources for instructors that have been shared by those who have adopted the laboratory manual for their courses. It includes institution-specific editions of the lab manual, instructor copies, assignments, and other ancillary and supporting materials. Instructors are granted read-only access to all files in the library, which allows them to view and download files, and to subsequently edit copies of those files for their own use. The read-only access also prevents unauthorized editing of other instructor's versions of the manual saved to the library. Instructors who develop their own editions of the manual are encouraged to share their editions and supporting materials to the community. We administer the library through Google Drive. Because of the sensitive nature of some course materials, access is limited to only instructors with proven credentials. To request access, please complete the following form: request instructor access to the Soils Laboratory Manual Library using this Google Form. Once you have been approved, we will contact you and provide access to the Google Drive folder for the Soils Laboratory Manual Library. If you do not have a Google-associated login, please contact Colby Moorberg.. He can make other arrangements to get you the materials you need.

New to Version 2

The original *Soils Laboratory Manual* was created in Microsoft Word. Version 2 was formatted in Pressbooks, an online book publishing platform. The manual is available through New Prairie Press as a PDF, EPUB and MOBI ebook formats, and as a web book. The web book version is adaptive to screen sizes, making it ideal for reading from small tablets or smart phones. Digital versions of the manual are also optimized for screen readers.

In addition to the format, the recommended readings from textbooks were replaced with free, online resources including extension publications, YouTube videos, and government reports and documents. This makes it possible for instructors that have adopted the *Soils Laboratory Manual* to make courses textbook-free.

The Soil Carbon and Respiration laboratory was completely re-imagined to focus more on the soil health aspects of soil respiration. In addition, the laboratory was made more financially accessible for institutions by eliminating the need to an expensive Infrared Gas Analyzer (IRGA) and instead measuring respiration using relatively inexpensive soil respiration kits from Solvita®. A similar activity was added to the Compost Facility Field Trip laboratory in which compost windrows are tested for maturity using compost kits from Solvita®. A virtual tour of the Kansas State Compost Facility was also added in the form of a YouTube video, facilitating use of this laboratory as a virtual laboratory for institutions that do not have easy access to industrial scale composting facilities. Lastly, the SoilWeb Field Trip laboratory was added, which is a self-led field trip that engages students with technology and soil mapping. This laboratory was created out of a need for social distanced activities during the COVID-19 pandemic.

The Instructor Copy

This instructor copy, currently available through the shared Google drive (see above), is designed to make teaching with the *Soils Laboratory Manual* easier. All answers that are blank in the student copy are completed in the instructor copy. These answers are italicized and in blue text. The instructor copy of Laboratory 12 – Soil Carbon and Respiration is in draft form at the moment. However, the instructor copy of that laboratory is expected to be finalized by summer of 2022. If you notice anything that needs a correction in the instructor copy, please notify Colby Moorberg.

Feedback and Questions

If you have corrections, comments, or questions about the *Soils Laboratory Manual*, please contact Colby Moorberg.

PART I

INTRODUCTION AND REVIEW

1. Safety Training and Orientation

Proper safety precautions are essential for any laboratory or fieldwork in soil science, and the first step towards ensuring safety for all students and instructors is to review the potential hazards, safety equipment, and procedures for the Soils Teaching Lab. In addition, a thorough understanding of the structure of this laboratory manual, and instructor expectations is essential for student success.

Learning Objectives

- Orient students on all safety aspects of the Soils Teaching Lab.
- Review the course syllabus and the structure of the Soils Laboratory Manual.

Materials

- Course syllabus

Course Syllabus

You will receive a course syllabus during lab and will review it with your classmates and instructor.

Lab Safety Training

- All students must exercise proper safety precautions in the laboratory. The safety guidelines below will be reviewed during the first lab period and as required during the semester.
- Food and drink are not allowed in the Agronomy 305 teaching laboratory.
- In the event of a fire alarm, the class will make an orderly exit of the laboratory, calmly proceed down the hall to the exit near the loading dock, and regroup on the loading dock parking lot.
- The emergency shower is on the east side of the laboratory near the bookshelf.
- Eyewashes are on the east side of the laboratory near the bookshelf.
- Sandals, flip-flops, and other open-toed shoes that do not fully cover the foot are not acceptable footwear for Agronomy 305 labs activities and field trips.
- Shorts, mini-skirts and other clothing that not fully covering the leg above the knee are not allowed for Agronomy

305 labs activities and field trips.

- Chemical spills should be immediately reported to the lab instructor.
- Broken glassware should be immediately reported to the lab instructor.
- Injuries during the laboratory activities should be immediately reported to the lab instructor.
- Transportation will be provided by K-State for all field trips.
- Some field trips may involve short hikes over uneven terrain. If you are unable to partake in such hikes, please notify the instructor at the beginning of the semester so that other arrangements can be made.
- Guests are not allow guests to join you during AGRON 305 laboratory sessions.

2. Chemistry and Math Review

Soil science, as with any science, involves collection and interpretation of data. In order to appropriately record and interpret data in this course, students must have a fundamental understanding of unit conversions and chemistry. This lab will review some of the major concepts that are essential for success in Soils.

Learning Objectives

- Review basic chemistry and math skills that will be used throughout the semester.

Materials

- Chemistry and Math for Soil Scientists Problem Set

Recommended Reading

Review relevant chemistry concepts in a chemistry textbook of your choosing.

Prelab Assignment

Using a chemistry textbook, and the conversion factors and formulas provided in this laboratory manual, consider the following questions.

1. Define dimensional analysis. Describe how it can be useful for unit conversion, and how it can be used to check the accuracy of calculations.
2. Define molarity and give several examples of how it can be expressed (labels or units).
3. Describe, in general terms, the process of an acid-base titration.

4. Note if any of the units listed in under Conversion Factors and Formulas are unfamiliar to you. If so, look that unit up and describe it in terms of units with which you are familiar.
5. Are the units "Mg" and "mg" the same? If not, which one is larger?
6. How many dimensions are there in the following units? Label each unit as a length, area, or volume.
   - m:
   - m^2:
   - m^3:

Introduction

Dimensional Analysis

Quite often, our measurements are not in the same units in which we wish to express our results. However, converting from one measurement to another is not difficult with the correct conversion factors. The key is appreciating that different units can be used to express the same amount. Calling a sofa by another name like couch does not change anything about that piece of furniture, nor does expressing a person's height in in instead of feet make them any taller or shorter. So by changing units, we do not change the amount, just the name we use to express that amount.

This is based on the principle that if we multiply any number by one, it does not change the number. For instance, we know that one hour is equal to 60 minutes, and that one minute is equal to 60 seconds. Consider the following fraction:

$$\frac{\text{60 minutes}}{\text{60 minutes}}$$

We know that this fraction reduces to one because the top (the numerator) is equal to the bottom (the denominator). However, we know that 60 minutes is equal to one hour. So we could write the fraction:

$$\frac{\text{60 minutes}}{\text{1 hour}}$$

While the numerical value of this fraction is 60, the amount of time represented in both the numerator and denominator are equal, and this fraction is, in a sense, equal to one. We use this concept, most likely without thinking about it, when we convert certain quantities in our head. If told that something will take about two hours, we can immediately convert this in our head to 120 minutes. Or if we need to be somewhere in half an hour, we can convert this to 30 minutes in our head without much effort. If we do think about it, however, we would recognize that the method used to make this conversion can be used to complete more complicated conversions that we cannot perform in our head. These conversions we performed in our head used the conversion factor 1 hour = 60 minutes and can be written as follows:

$$\frac{\text{2 hours}}{1} \times \frac{\text{60 minutes}}{\text{1 hour}} = \text{120 minutes}$$

In this case, when we write it out, the difference is that we flip the conversion factor so that 1 hour is in the numerator and 60 minutes is in the denominator. The most important aspect dimensional analysis is to always keep track of your units. By keeping track of your units, you make sure units cancel properly. In the previous example, the conversion factor has minutes in the denominator to cancel the minutes in the initial value; hours are in the numerator to take the place of the minutes.

$$\frac{120 \text{ minutes}}{1} \times \frac{1 \text{ hour}}{60 \text{ minutes}} = 2 \text{ hours}$$

If this is set up incorrectly, the conversion would yield:

$$\frac{120 \text{ minutes}}{1} \times \frac{60 \text{ minutes}}{1 \text{ hour}} = \frac{7,200 \text{ minutes}^2}{\text{hour}}$$

This answer is not only incorrect, but it makes very little sense.

Chemical Titrations

Reactions in aqueous solution are commonly used to determine the unknown concentration of a dissolved substance. Titration is the technique used:

- A solution of known concentration (standard solution) is added quantitatively to a known volume of the solution of unknown concentration. For example, a burette is used to slowly add a solution with a known concentration of HCl to a specific volume of a solution with an unknown concentration of NaOH.
- The HCl and NaOH react to form NaCl and H_2O. The point in the titration at which the last of the NaOH reacts with the added HCl is called the *equivalence point*. This point is commonly detected using an indicator that changes color at the equivalence point. Thus, you know to stop titrating when the color changes.
- Knowing the concentration and volume of the added HCl solution and the original volume of the NaOH solution allows us to calculate the concentration of NaOH in the original solution.
  - We can calculate the number of moles of HCl in the solution using the following:

$$\text{Moles HCl} = (\text{volume HCl added}) \times (\text{molarity of HCl})$$

  - We know that when HCl and NaOH react, the equations is: HCl + NaOH → NaCl + H_2O. Thus, one mole of HCl reacts with one mole of NaOH. So if we know the number of moles of HCl required for the reaction, we can calculate the number of moles of NaOH that reacted.
- Note: One practical application of titration is in determining blood alcohol content. Blood plasma is titrated with potassium dichromate, which oxidizes the alcohol. The amount of potassium dichromate required to do so is used to calculate the amount of alcohol present.

Let's work an example:

We would like to determine the molarity of a NaOH solution of unknown concentration. Of this solution, 50 mL is titrated with 0.01 molar HCl solution. Of the HCl solution, 20 mL are required to reach the equivalence point (sometime called the *end point*).

1. Calculate the moles of HCl used in the titration:

$$\text{Moles of HCl} = \frac{20\text{ mL HCl}}{1} \times \frac{0.01\text{ moles HCl}}{1\text{ L HCl}} \times \frac{1\text{ L HCl}}{1000\text{ mL HCl}} = 2 \times 10^{-4}\text{ moles HCl}$$

2. Calculate the moles of NaOH present:

Since we know that one mole of NaOH reacts with one mole of HCl, we know that 2 x 10^{-4} moles of NaOH are present in the solution.

$$\text{Moles of NaOH} = \frac{2 \times 10^{-4}\text{ moles HCl}}{1} \times \frac{1\text{ mole NaOH}}{1\text{ mole HCl}} = 2 \times 10^{-4}\text{ moles NaOH}$$

3. Calculate the molarity of the NaOH:

The original volume of the NaOH solution was 50 mL. So 50 mL of solution contains 2 x 10^{-4} moles of NaOH. Therefore, the molarity of NaOH is:

$$\frac{2 \times 10^{-4}\text{ moles NaOH}}{50\text{ mL NaOH}} \times \frac{1000\text{ mL NaOH}}{1\text{ L NaOH}} = \frac{0.004\text{ moles NaOH}}{\text{L}}\text{ or } 0.004\text{ M NaOH}$$

Let's work another example:

We would like to determine the molarity of a $Ca(OH)_2$ solution of unknown concentration. Of this solution, 80 mL is titrated with 0.01 molar HCl solution. Of the HCl solution, 35 mL are required to reach the equivalence point. First, let's write the equation for this reaction:

$2HCl + Ca(OH)_2\ CaCl_2 + 2H_2O$

1. Calculate the moles of HCl used in the titration.

$$\text{Moles of HCl} = \frac{35\text{ mL HCl}}{1} \times \frac{0.01\text{ moles HCl}}{1\text{ L HCl}} \times \frac{1\text{ L HCl}}{1000\text{ mL HCl}} = 3.5 \times 10^{-4}\text{ moles HCl}$$

2. Calculate the moles of $Ca(OH)_2$ present:

In contrast to the reaction between HCl and NaOH, we see that two moles of HCl are required to react with one mole of $Ca(OH)_2$. Therefore, the moles of $Ca(OH)_2$ in solution are:

$$\frac{3.5 \times 10^{-4}\text{ moles HCl}}{1} \times \frac{1\text{ mole Ca(OH)}_2}{2\text{ moles HCl}} = 1.75 \times 10^{-4}\text{ moles Ca(OH)}_2$$

- Calculate molarity of the $Ca(OH)_2$

The original volume of the $Ca(OH)_2$ solution was 80 mL. So 80 mL of solution contains 1.75 x 10-4 moles $Ca(OH)_2$. Therefore, the molarity is

$$\frac{1.75 \times 10^{-4}\text{ moles Ca(OH)}_2}{80\text{ mL Ca(OH)}_2} \times \frac{1000\text{ mL Ca(OH)}_2}{1\text{ L Ca(OH)}_2} = \frac{2.19 \times 10^{-3}\text{ moles Ca(OH)}_2}{\text{L}}\text{ or } 2.19 \times 10^{-3}\text{ M Ca(OH)}_2$$

Activity 1 & Assignment: Problem Set

A problem set will be provided to you at the beginning of the lab. You will work on the problem set during the laboratory period, and turn it in as instructed.

PART II

SOIL FORMATION, CLASSIFICATION, AND MAPPING

3. Soil Formation

Soils are incredibly diverse in their appearance, form, function, and use throughout the globe. A fundamental understanding of soil formation allows us to more appropriately manage those soil resources. For example, the Flint Hills that surround Manhattan, Kansas encompass some of the last remaining tallgrass prairie in the U.S., primarily due to the soil. The rocky outcroppings, steep slopes, and boulder-ridden soil profiles make tillage of the soils for row crop production nearly impossible. However, the landscape lends itself well to grazing, allowing the region to be managed as rangeland instead of wheat fields. However, at a finer scale, land uses other than rangeland do exist in the region; and soil mapping allows us to identify those soils suitable for such land uses. The concepts of soil formation covered in this lab provide the basis for soil mapping, which is covered in more detail in Soil Classification and Mapping.

Learning Objectives

- Recognize common rocks and minerals.
- Relate rock and mineral composition to (a) ease of weathering and (b) products of weathering.
- Understand the general geology of Kansas.
- Know the major types of soil parent materials.
- Know the five factors of soil formation.
- Recognize major soil differences as they are affected by differences in soil formation.

Materials

- Mineral collection (Item #GEO2194, Carolina Biological Supply Company, Burlington, North Carolina, U.S.)
- Igneous, metamorphic, and sedimentary rock collections (Item # GEO115, Carolina Biological Supply Company, Burlington, North Carolina, U.S.)
- Rock cycle chart (Item #GEO8658, Carolina Biological Supply Company, Burlington, North Carolina, U.S.)
- Generalized Geologic Map of Kansas (Kansas Geological Survey, Lawrence, Kansas, U.S.)
- Parent material collection (Various sources and peer instructors from around the U.S.)
- Soil monoliths (Colby, Ulysses, Tivoli, Cherokee, and Kenoma soil series)

Recommended Reading

- Weathering and Soil (Earle, 2015)
- Soil Formation and Classification (USDA NRCS, 2020)

Prelab Assignment

Read the recommended reading resources and the introduction to this lab, then answer consider the questions listed below. These definitions/questions will provide a concise summary of the major concepts addressed in the lab. They are also useful as study notes for exams.

- Define and Explain the differences between igneous rock, sedimentary rock, and metamorphic rock.
- Define and give a specific example of each of the following terms:
    - Physical weathering
    - Chemical weathering
    - Primary mineral
    - Secondary mineral
- Define the following three parent materials:
    - Alluvium
    - Colluvium
    - Residuum
- What are the five factors of soil formation?

Introduction

Understanding soils and why they are different requires knowing how soil forms. Soils are the product of weathering of pre-existing parent materials, so knowing rocks and minerals is important to understanding differences among soils.

Minerals

Minerals are naturally occurring inorganic substances with characteristic composition and physical properties. Based on their origin, minerals are classified as primary or secondary. Primary minerals form during cooling and solidification of the original molten material of the earth's crust. Secondary minerals form from the decomposition products of primary minerals.

Rocks

Rocks are combinations of two or more minerals. The three major groups of rocks are igneous, sedimentary, and metamorphic. Igneous rocks form as molten mineral matter cools and solidifies. They are the most abundant group of rocks in the earth's crust. Sedimentary rocks originate from weathering of igneous rocks. Sedimentary rocks are the most abundant rock type on the earth's surface. Metamorphic rocks form when igneous or sedimentary rocks are modified by heat and/or pressure.

Igneous rocks

Types of igneous rocks, their mineral composition, and texture (grain size) are shown in Figure 3.1. The vertical arrangement of rocks in the figure is based on texture, from coarse to fine. The texture is dictated by the rate of cooling of the molten magma from which the rocks form. When the magma cools slowly, far below the earth's surface, sufficient time is available for relatively large mineral crystals to form. In contrast, when magma cools rapidly at the earth's surface, minimal time is available for crystallization, and the crystals are small. Thus, rocks forming deep within the earth (intrusive rocks like granite) are coarse textured, while rocks forming at the surface (extrusive rocks like obsidian) are fine.

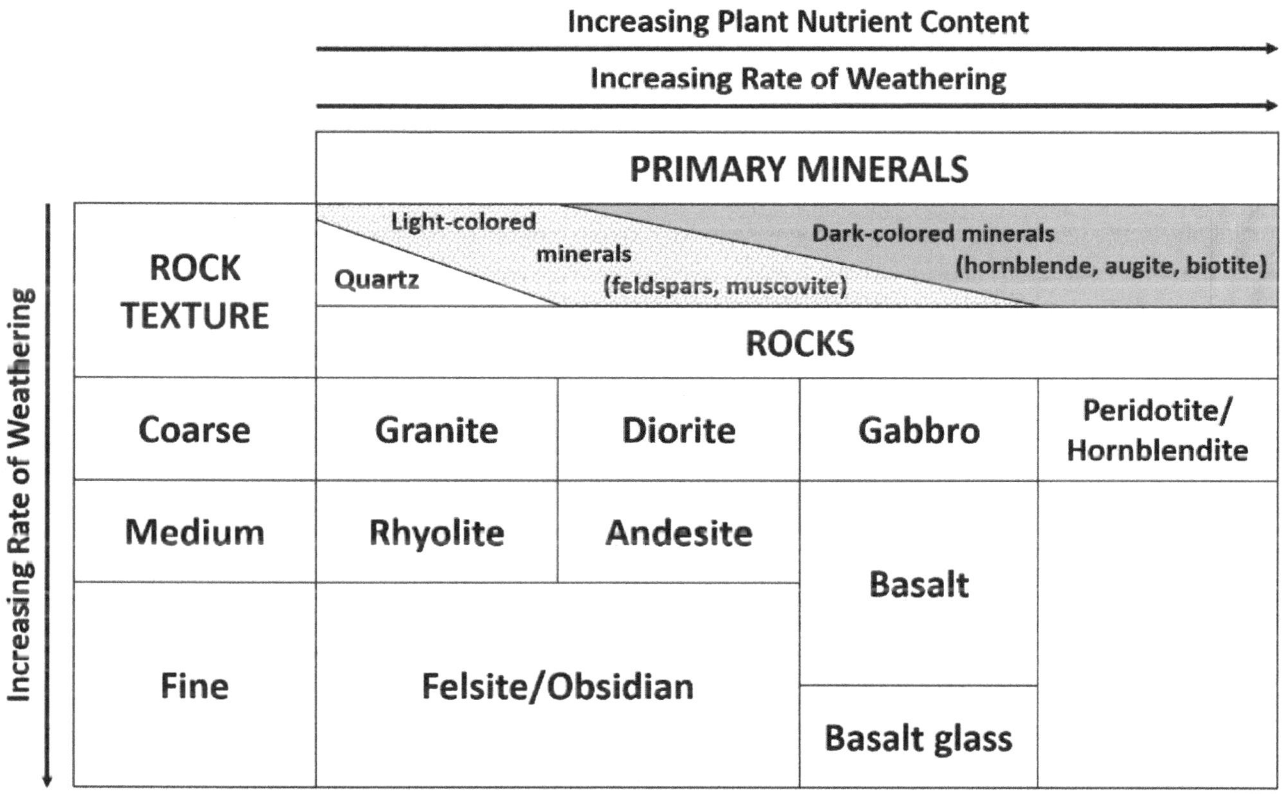

Figure 3.1. Characteristics of igneous rocks. Diagram adapted from Brady and Weil (2009) wand King et al. (2003) with permission.

The horizontal arrangement in Figure 3.1 is based on mineral composition. For example, granite is composed mostly of quartz and light-colored minerals like feldspars and muscovite. In contrast, peridotite is composed mostly of dark-colored minerals (hornblende, augite, and biotite).

Both the mineral makeup of igneous rocks and their texture are important in soil formation. Minerals vary in solubility. Biotite ($K(Mg,Fe)_3AlSi_3O_{10}(OH)_2$) contains more Mg and Fe, and is more soluble than orthoclase feldspar ($KAlSi_3O_8$), which itself is more soluble than quartz (SiO_3). Texture affects the amount of surface area exposed to weathering; that is, small crystals have a large specific surface area (surface area/unit of weight) and thus dissolve faster than large crystals. Therefore, from Figure 3.1, we can generalize that the rate of weathering of igneous rocks increases from the upper left corner (granite) to the lower right corner; in other words, as grain size decreases, solubility increases.

Sedimentary rocks

Sedimentary rocks are classified as clastics or precipitates. Clastics are rocks formed through physical means: fragments of igneous rocks or sediments derived from igneous rocks are cemented together to form sedimentary rocks. The cementing agents include clay, iron oxide, silica, or calcium carbonate. Texture of the resulting rocks can vary.

Table 3.1 Rock Texture

Texture	Examples	Particle Diameter, mm
Coarse	Conglomerates	>2 (pebbles to boulders)
Medium	Sandstone	0.05 to 2.00 (sand size)
Fine	Shale	<0.005 (silt/clay)

Table courtesy of King et al. (2003).

Precipitate-type sedimentary rocks form as ions from solution precipitate either chemically or biochemically. Biochemical precipitates usually involve marine animals and may contain fossils and shells. Limestone is an example of a precipitate sedimentary rock.

Metamorphic rocks

One way to explain metamorphic rocks is the types of igneous or sedimentary rocks from which they were formed. Table 3.2 provides some examples.

Table 3.2. Metamorphic rock formation

Original Rock	Metamorphic Rock
Granite	Gneiss
Shale	Slate
Sandstone	Quartzite
Limestone	Marble

A summary of metamorphic rock examples and the original rocks from which the metamorphic rocks were made. Table courtesy of King et al. (2003).

Activity 1: Elemental composition of minerals

Table 3.3 shows several primary minerals (formed as magma cools) and secondary minerals (formed by weathering or recrystallization of weathering products, i.e., in sedimentary rocks), their composition, and their resistance to weathering.

Table 3.3. Properties of minerals.

Primary Mineral†	**Composition**	**Secondary Mineral‡**	**Composition**	**Weathering**
		Goethite	FeOOH	Most resistant
		Hematite	Fe2O3	↓
		Gibbsite	Al2O3•3H2O	↓
Quartz	SiO2			↓
		Clay minerals	Al silicates	↓
Muscovite	KAl3Si3O10(OH)2			↓
Microcline	KAlSi3O8			↓
Orthoclase	KAlSi3O8			↓
Biotite	KAl(Mg,Fe)3Si3O10(OH)2			↓
Albite	NaAlSi3O8			↓
Hornblende	Ca2Al2Mg2Fe3Si6O22(OH)2			↓
Augite	Ca2(Al,Fe)4(Mg,Fe)4Si6O24			↓
Anorthite	CaAl2Si2O8			↓
Olivine	(Mg,Fe)2SiO4			↓
		Dolomite	CaCO3•MgCO3	↓
		Calcite	CaCO3	↓
		Gypsum	CaSO4•2H2O	Least resistant

†Primary minerals are found in igneous rocks or metamorphosed igneous rocks.‡Secondary minerals are most commonly found in sedimentary rocks. §Table courtesy of King et al. (2003).

Using the data in Table 3.4, put a checkmark in each column box where a mineral contains the nutrient for that column. Hornblende is completed as an example.

Table 3.4. Essential plant nutrients found in primary and secondary minerals.

	N	P	K	Ca	Mg	S	Fe	Mn	Cu	Zn
Primary Minerals										
Quartz										
Muscovite										
Orthoclase										
Biotite										
Hornblende				√	√		√			
Olivine										
Augite										
Albite										
Secondary Minerals										
Goethite										
Gibbsite										
Dolomite										
Calcite										
Gypsum										

The soil based plant nutrients are listed below.

Primary Macronutrients

- Nitrogen (N)
- Phosphorus (P)
- Potassium (K)

Secondary Macronutrients

- Calcium (Ca)
- Magnesium (Mg)
- Sulfur (S)

Micronutrients

- Iron (Fe)
- Manganese (Mn)
- Boron (B)
- Zinc (Zn)
- Copper (Cu)
- Chlorine (Cl)
- Molybdenum (Mo)
- Nickel (Ni)

Which mineral(s) in Table 3.3 will result in the formation of the most fertile soil?

Which mineral(s) in Table 3.3 will result in the formation of the least fertile soil?

Activity 2: Mineral and Rock Investigation

Examine the rock and mineral specimens on display. Note differences in color, color pattern, shape, and the cleavage angles on the mineral specimens. For the rock examples, note the differences in the three broad classes of rocks: igneous, metamorphic, and sedimentary. Study the size of the individual minerals contained in the rocks. Examine the color patterns and the arrangement of the minerals. Try to identify specific minerals contained in the rocks.

As you study the rock samples, refer to the Table 3.5. Note the differences in mineral size (fine-texture versus coarse-texture), the differences in mineral content between the dark (basic or mafic) and the light (acid or felsic), and the elemental content of the igneous rock types.

As you review the rock and mineral types, complete Table 3.5 and Table 3.6. These tables will provide a useful summary of properties of the major rocks and minerals important in soil formation.

Table 3.5. Summary of important soil minerals.

Mineral	Color	Plant Nutrients	Relative Rate of Weathering
Quartz			
Muscovite mica			
Biotite mica			
Hornblende			
Calcite			
Dolomite			

Table courtesy of King et al. (2003).

Table 3.6. Summary of important soil forming rocks.

Rock	Class	Grain size	Dominant minerals	Rate of weathering	Potential nutrient supply
	Igneous, metamorphic, or sedimentary	Coarse or fine		Fast, moderate, or slow	High, medium, or low
Granite					
Gabbro	Igneous	Coarse	Feldspars, muscovite, hornblende, augite	Moderate	High
Gneiss					
Schist	Metamorphic	Coarse	Mica, quartz, hornblende	Moderate	Medium
Quartzite					
Slate					
Limestone					
Sandstone					
Shale					

Table courtesy of King (2003).

Activity 3: Rock Cycle Model

The rock cycle model is two dimensional and shows a cross-section through the earth's crust. Vertically, it represents a depth of two to five miles. Horizontally, it could represent distances ranging from a few miles to several hundred miles.

Examine first the rock cycle showing the relationships between igneous, sedimentary, and metamorphic rocks. Note the processes/factors (heat, pressure, erosion, etc.) that change one rock class to another.

Shale can be transformed into shist through metamorphism. What metamorphic rock acts as an intermediate?

What process is responsible for the formation of layers, a characteristic of sedimentary rocks?

What two forces can cause quartz in a sandstone rock to be changed to metamorphic quartzite?

Activity 4: Geology of Kansas

In this exercise, you will be using the geologic map of Kansas on the next page. This map contains much information on different sedimentary materials, as well as glacial and wind-blown sediment. Review the map, and answer the following questions.

Of the three different major rock classes (igneous, metamorphic, sedimentary), only one is found in Kansas. Name the rock class that is found in Kansas. Also, name one rock formation from the map, and describe where in the state that rock formation can be observed.

What is the oldest deposit exposed on the surface in Kansas? How old is it, and where in the state does it surface?

Describe the pattern occurring between sediment age and its order among other sediment layers from most deep to most shallow.

Where have glaciers deposited material in Kansas?

What sedimentary deposit(s) underlies Manhattan, KS? How old is it?

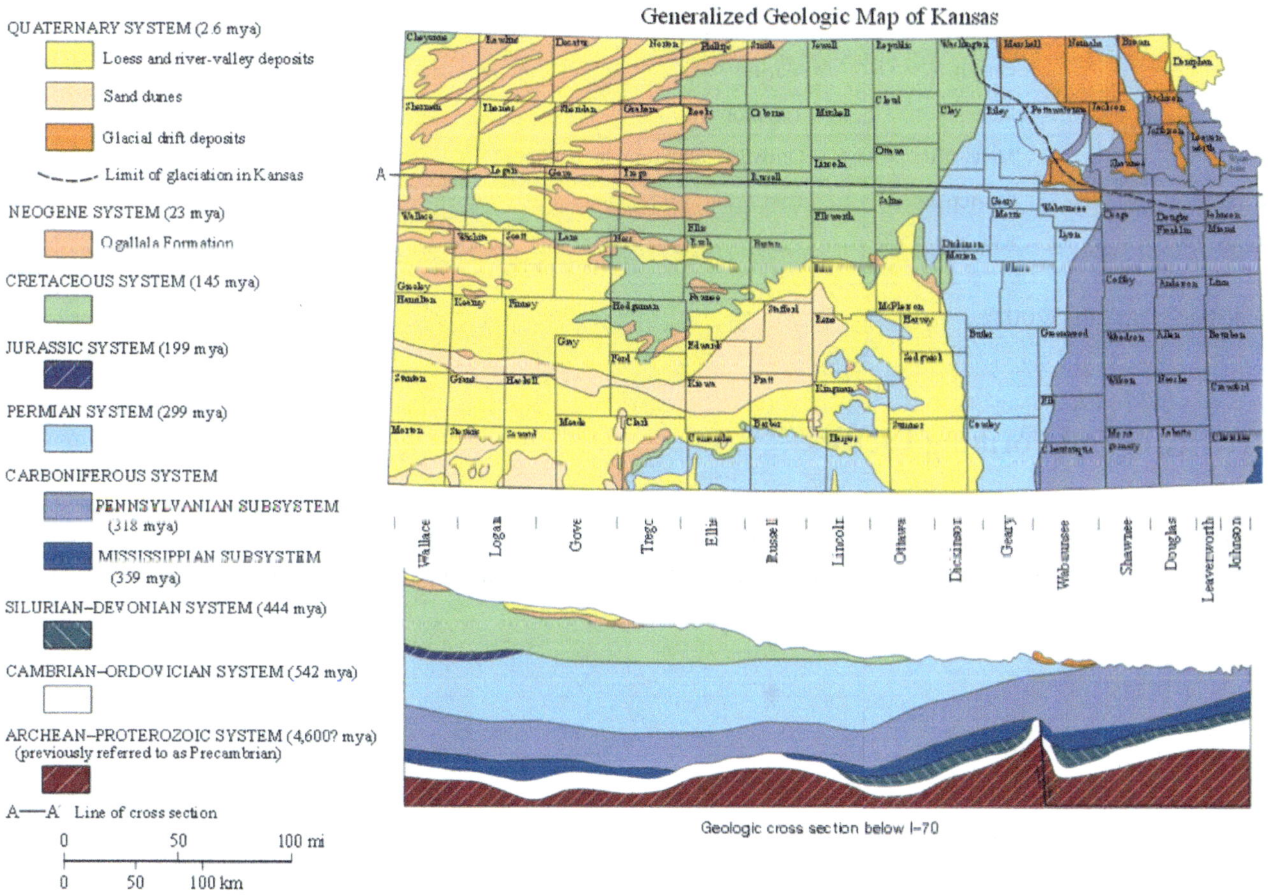

Figure 3.2. Generalized Geologic Map of Kansas. Map courtesy of the Kansas Geological Survey, used with permission.

Activity 5: Types of Soil Parent Materials

All mineral soils develop from the rocks and minerals of the earth's crust. However, the original rock material commonly is transported or reworked by some mechanism before a soil is formed. Variations in the history of the initial material lead to a range of types of soil parent materials. Table 3.7 summarizes various categories of soil parent materials:

Describe the various types of parent materials on display.

Table 3.7. Summary of transported parent materials.

Mode of transportation	Mode of deposition	Name of parent material
Gravity	Gravity	Colluvium
Water	Stream	Alluvium
Water	Lake	Lacustrine
Water	Ocean	Marine
Ice	Ice	Till, moraine
Ice	Meltwater	Outwash (alluvium, lacustrine)
Wind	Suspension	Loess
Wind	Saltation, Creep	Dunes

Table based on King (2003).

Activity 6: Soil Forming Factors

While the soil parent material is important to the type of soil that may form, other forces or factors also influence weathering and soil formation. The resulting soil properties we observe are the result of the interaction of these factors.

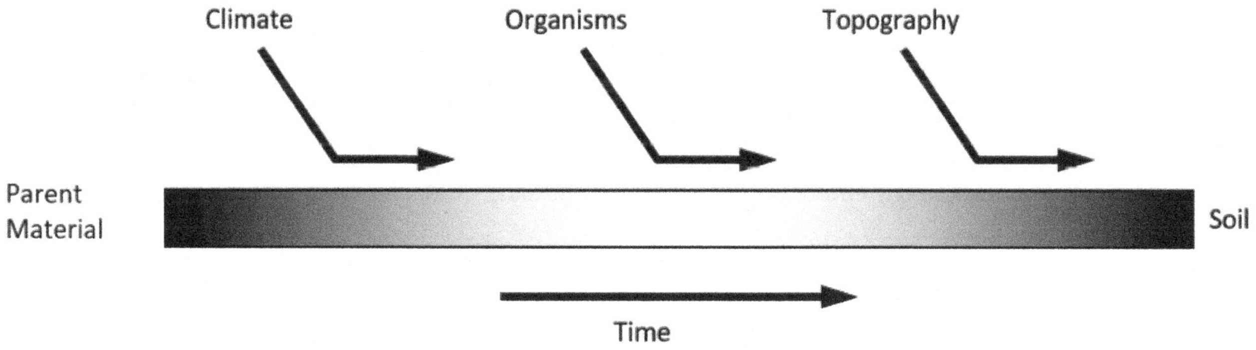

Figure 3.3. The soil forming factor continuum. Diagram from King et al. (2003).

As Figure 3.3 suggests, soil formation is a continuum. As any one factor or influence changes, the final soil product will vary. An infinite number of combinations of soil forming processes yield many different soil properties. Understanding the role of each soil forming factor allows us to understand or predict soil occurrence.

Study the soil monoliths (profiles) and note the differences in color, depth, and horizonation. Review the major soil forming factors for each as a means of explaining the differences you observe.

After studying the monoliths on display and their accompanying Official Series Descriptions, name the soil-forming factor(s) most responsible for the differences in the following soils. (There may be more than one.)

Table 3.8. Identifying the most prevelant soil forming factor

Soils	Main soil-forming factor(s) causing difference:
Example: Norfolk versus Cecil	Parent Material: Norfolk = Coastal Plain sediments Cecil = granite or gneiss
Colby versus Ulysses	
Ulysses versus Tivoli	
Cherokee versus Kenoma	

COLBY SERIES

The Colby series consists of very deep, well drained and somewhat excessively drained, moderately permeable soils that formed in loess. Colby soils are found on plains and hillslopes of tableland in the Central High Tableland. Slopes range from 0 to 60%. Mean annual precipitation is about 450 millimeters (18 in), and the mean annual temperature is 11°C (52°F).

TAXONOMIC CLASS: Fine-silty, mixed, superactive, calcareous, mesic Aridic Ustorthents

TYPICAL PEDON: Colby silt loam (Colors are for dry soils.)

A—0 to 10 cm (0 to 4 in); grayish brown (10YR 5/2) silt loam, dark grayish brown (10YR 4/2) moist; weak fine platy structure to a depth of 5 cm (2 in), weak fine granular structure below 5 cm (2 in); slightly hard, friable; strong effervescence; moderately alkaline; gradual smooth boundary. (8 to 15 cm (3 to 6 in) thick)

AC—10 to 20 cm (4 to 8 in); light brownish gray (10YR 6/2) silt loam, grayish brown (10YR 5/2) moist; weak fine granular structure; slightly hard, friable; strong effervescence; moderately alkaline; gradual smooth boundary. (0 to 23 cm (0 to 9 in) thick)

C1—20 to 51 cm (8 to 20 in); pale brown (10YR 6/3) silt loam, brown (10YR 5/3) moist; massive; slightly hard, friable; few fine roots and root channels; porous; few soft lime accumulations; violent effervescence; moderately alkaline; gradual wavy boundary. (25 to 76 cm (10 to 30 in) thick)

C2—51 to 200 cm (20 to 79 in); very pale brown (10YR 7/3) silt loam, light yellowish brown (10YR 6/4) moist; massive; slightly hard, friable; porous; violent effervescence; moderately alkaline.

GEOGRAPHIC SETTING: The Colby soil series forms in loess on nearly level to steep hillslopes and uplands in a tableland landscape. Slopes are commonly 3 to 15%, but range from 0 to 60%. Mean annual air temperatures ranges from 7 to 13°C (45 to 55°F), and mean annual precipitation ranges from 33 to 51 cm (13 to 20 in).

DRAINAGE AND SATURATED HYDRAULIC CONDUCTIVITY:
The Colby soil series is well drained to somewhat excessively drained. Runoff is negligible to high. Saturated hydraulic conductivity is moderately high.

DISTRIBUTION AND EXTENT: Western Kansas, eastern Colorado, western Nebraska, southwestern South Dakota, and eastern Montana in MLRA 64, 67, and 72. The series is extensive.

SERIES ESTABLISHED: Reconnaissance Soil Survey of Western Kansas, 1910.

[Adapted from National Cooperative Soil Survey (2012)]

ULYSSES SERIES

The Ulysses series consists of very deep, well drained soils that formed in loess. Ulysses soils are on hillslopes and plains of tableland in the Central High Tableland (MLRA 72). Slopes range from 0 to 20%. Mean annual precipitation is 455 mm (18 in), and the mean annual temperature is 12°C (54°F).

TAXONOMIC CLASS: Fine-silty, mixed, superactive, mesic Aridic Haplustolls

TYPICAL PEDON: Ulysses silt loam (Colors are for dry soil unless otherwise noted.)

Ap—0 to 10 cm; (0 to 4 in); grayish brown (10YR 5/2) silt loam, very dark grayish brown (10YR 3/2) moist; weak fine granular structure; hard, friable; neutral; abrupt smooth boundary.

A—10 to 25 cm; (4 to 10 in); dark grayish brown (10YR 4/2) silt loam, very dark grayish brown (10YR 3/2) moist; moderate medium granular structure; hard, friable; many worm casts; moderately alkaline; gradual smooth boundary. (The combined thickness of the A horizon is 18 to 30 cm (7 to 12 in).)

Bw—25 to 46 cm; (10 to 18 in); grayish brown (10YR 5/2) silt loam, dark grayish brown (10YR 4/2) moist; moderate medium granular structure; hard, friable; many worm casts; strong effervescence; moderately alkaline; gradual smooth boundary.

C1—46 to 76 cm; (18 to 30 in); pale brown (10YR 6/3) silt loam, brown (10YR 5/3) moist; massive; slightly hard, very friable; violent effervescence; faint films and streaks of segregated lime; moderately alkaline; gradual smooth boundary.

C2—76 to 200 cm; (30 to 79 in); pale brown (10YR 6/3) silt loam, brown (10YR 5/3) moist; massive; soft, very friable; strong effervescence; moderately alkaline. (The combined thickness of the C horizon is 38 to 102 cm (15 to 40 in))

GEOGRAPHIC SETTING: The Ulysses soil series forms in loess on hillslopes and plains on tableland. Slopes are commonly 1 to 8%, but range from 0 to 20%. Mean annual air temperature ranges from 8 to 14°C (46 to 57°F). Mean annual precipitation ranges from 35 to 64 cm (14 to 25 in). Most of the rainfall occurs as high-intensity, convective thunderstorms from late spring through early autumn. Precipitation in winter generally occurs as snow. The frost-free period is 135 to 210 days. The Thornthwaite Annual P-E Index ranges from 25 to 34.

DRAINAGE AND SATURATED HYDRAULIC CONDUCTIVITY:
The Ulysses series is well drained, with low or medium runoff, and moderately high saturated hydraulic conductivity.

DISTRIBUTION AND EXTENT: Western Kansas, western Nebraska, southeastern Wyoming, southwestern South Dakota, eastern Colorado, and Oklahoma Panhandle. LRR H, Major Land Resource Area 72; Central High Tableland. The series is of large extent.

SERIES ESTABLISHED: Logan County, Kansas, 1957.

[Adapted from National Cooperative Soil Survey (2017)]

TIVOLI SERIES

The Tivoli series consists of very deep, excessively drained, rapidly permeable soils that formed in sandy eolian sediments. These soils occur on undulating to hummocky sand dunes on stream terraces. Slope is complex and ranges from 1 to 45%. Mean annual precipitation is 660 mm (26 in), and the mean annual temperature is 16.1°C (61°F).

TAXONOMIC CLASS: Mixed, thermic Typic Ustipsamments

TYPICAL PEDON: Tivoli fine sand, rangeland.
(Colors are for dry soil unless otherwise stated.)

A—0 to 18 cm (0 to 7 in); pale brown (10YR 6/3) fine sand, brown (10YR 4/3) moist; single grained, loose, very friable; many roots; neutral; gradual smooth boundary. Thickness is 10 to 25 cm (4 to 10 in)

C—18 to 152 cm (7 to 60 in); yellow (10YR 7/6) fine sand, brownish yellow (10YR 6/6) moist; single grained, loose; roots decrease as depth increases; neutral.

GEOGRAPHIC SETTING: The Tivoli soil series forms in sandy eolian sediments in river valleys and dissected plains. The series forms on sand dunes on sand sheets and stream terraces.

Slopes range from 1 to 45%. Mean annual precipitation franges from 51 cm to 81 cm (20 to 32 in). Mean annual air temperature ranges from 13.9 to 17.8°C (57 to 64°F). The Thornthwaite annual P-E index values ranges from 30 to 44. The frost-free period is 185 to 240 days. The elevation ranges from 304.8 to 884 m (1000 to 2,900 ft).

DRAINAGE AND PERMEABILITY: Excessively drained. Permeability is rapid. Runoff is negligible on 1 to 5% slopes, very low on 5 to 20% slopes, and low on slopes greater than 20%.

DISTRIBUTION AND EXTENT: Central Rolling Red Plains, Western and Eastern Parts (MLRAs 78B and 78C) of Oklahoma, Kansas, and Texas. The series is extensive.

SERIES ESTABLISHED: Major County, Oklahoma; 1936.

[Adapted from National Cooperative Soil Survey (2016)]

CHEROKEE SERIES

The Cherokee Series is very deep, somewhat poorly drained soils that formed in fine textured sediments of the Cherokee Prairies. Slope ranges from 0 to 3%. The mean annual precipitation is 981 mm (40 in), and the mean annual air temperature is 16.1°C (61°F).

TAXONOMIC CLASS: Fine, mixed, active, thermic Typic Albaqualfs

TYPICAL PEDON: Cherokee silt loam cultivated. (Colors are for moist soils unless otherwise stated.)

Ap—O to 6 in; dark grayish brown (10YR 4/2), crushed silt loam; weak medium granular structure; soft, very friable; common very fine and fine roots throughout; common very fine and fine moderate continuity tubular pores; very strongly acid; clear smooth boundary. (4 to 10 in thick)

Al—6 to 12 in; dark grayish brown (10YR 4/2) crushed, silt loam; common fine distinct olive brown (2.5Y 4/4) moist irregular mottles throughout; moderate coarse platy structure parting to moderate coarse granular; soft, very friable; common very fine and fine roots throughout; common very fine and fine moderate continuity tubular pores and common coarse moderate continuity tubular pores; very strongly acid; clear smooth boundary. (0 to 6 in thick)

E—12 to 15 in; dark grayish brown (10YR 4/2) crushed, silt loam; common fine distinct dark yellowish brown (10YR 4/4) moist irregular mottles throughout; weak medium subangular blocky structure parting to weak medium granular; soft, very friable; common very fine and fine roots throughout; common very fine and fine moderate continuity tubular pores; moderately acid; abrupt smooth boundary. (3 to 8 in thick)

2Btgl—15 to 26 in; very dark grayish brown (10YR 3/2) exterior, silty clay loam; common medium distinct yellowish brown (10YR 5/6) moist irregular mottles throughout; moderate medium prismatic structure parting to strong medium angular blocky; extremely hard, very firm; common very fine roots between peds; common very fine low continuity tubular pores; many faint continuous very dark grayish brown (10YR 3/2), moist, clay films on vertical and horizontal faces of peds; few fine rounded black (N 2/0) hard iron-manganese concretions pedogenic throughout; strongly acid; abrupt smooth boundary. (5 to 15 in thick)

3Btg2—26 to 31 in; dark grayish brown (10YR 4/2) crushed, silty clay; common coarse prominent dark yellowish brown (10YR 4/6) moist irregular mottles throughout; moderate medium prismatic structure parting to moderate coarse subangular blocky; extremely hard, very firm; common very fine and fine roots throughout; common very fine moderate continuity tubular pores; common distinct continuous very dark grayish brown (10YR 3/2), moist, clay films on vertical and horizontal faces of peds; strongly acid; clear wavy boundary. (0 to 15 in thick)

3Btg3—31 to 35 in; grayish brown (10YR 5/2) crushed, silty clay; common fine prominent dark yellowish brown (10YR 4/6) moist irregular mottles throughout; moderate medium subangular blocky structure; very hard, friable; common very fine and fine roots throughout; common very fine moderate continuity tubular pores; few faint continuous dark grayish brown (10YR 4/2), moist, clay films on vertical and horizontal faces of peds; common medium rounded very dark grayish brown (10YR 3/2) soft clay bodies pedogenic throughout and few medium rounded white (10YR 8/1) soft nests of gypsum pedogenic throughout; strongly acid; clear wavy boundary. (0 to 10 in thick)

3Btg4—35 to 42 in; grayish brown (10YR 5/2) crushed, silty clay loam; common fine prominent dark yellowish brown (10YR 4/6) moist irregular mottles throughout and common fine prominent yellowish brown (10YR 5/8) moist irregular mottles throughout; weak coarse prismatic structure parting to moderate medium subangular blocky; very hard, friable; common very fine and fine roots throughout; common very fine and fine moderate continuity tubular pores; few faint discontinuous dark grayish brown (10YR 4/2), moist, clay films on vertical faces of peds and few distinct continuous very dark grayish brown (10YR 3/2) moist, silt coats in root channels and/or pores; few medium rounded white (10YR 8/1) soft nests of gypsum pedogenic throughout; strongly acid; gradual wavy boundary. (0 to 10 in thick)

3Btg5/E—42 to 54 in; dark grayish brown (10YR 4/2) exterior, silty clay loam; common fine prominent dark yellowish brown (10YR 4/6) moist irregular mottles throughout; moderate coarse prismatic structure parting to moderate medium subangular blocky; hard, friable; common very and fine roots throughout; many very fine and fine moderate continuity tubular pores; few faint discontinuous dark grayish brown (10YR 4/2) moist clay films on vertical faces of peds and few distinct continuous very dark grayish brown (10YR 3/2) moist, silt coats in root channels and/or pores; strongly acid; gradual wavy boundary. (0 to 15 in thick)

3Btg6/E—54 to 69 in; light brownish gray (10YR 6/2) exterior, silty clay loam; common coarse prominent dark yellowish brown (10YR 4/6) moist irregular mottles throughout and common coarse prominent dark reddish brown (5YR 3/4) moist irregular mottles throughout; moderate coarse prismatic structure parting to weak coarse subangular blocky; slightly hard, very friable; common very fine and fine roots throughout; many very fine and fine moderate continuity tubular pores; few faint continuous grayish brown (10YR 5/2), moist, clay films on vertical faces of peds and few distinct continuous dark grayish brown(10YR 4/2) moist silt coats in root channels and/or pores; common fine rounded black (N 2/0) soft iron/manganese concretions pedogenic throughout; strongly acid; gradual wavy boundary. (0 to 20 in thick)

3Btg7/E—69 to 86 in; light brownish gray (10YR 6/2) exterior, silty clay loam; common coarse prominent dark yellowish brown (10YR 4/6) moist irregular mottles throughout and common fine prominent dark brown (7.5YR 3/4.) moist irregular mottles throughout; weak coarse prismatic structure; slightly hard, very friable; few very fine roots throughout; common very fine and fine moderate continuity tubular pores; few faint discontinuous dark grayish brown (10YR 4/2), moist, clay films on vertical faces of peds and few distinct continuous grayish brown (10YR 5/2) moist, silt coats in root channels and/or pores; common fine rounded black (N 2/0) soft iron/manganese concretions pedogenic throughout and few medium irregular brown (10YR 5/3) soft clay bodies pedogenic in cracks; very strongly acid; clear smooth boundary. (0 to 20 in thick)

3Btg8/E—86 to 97 in; 80% light brownish gray (2.5Y 6/2) exterior, and 15% yellowish brown (10YR 5/8), exterior, and 5% strong brown (7.5 YR 4/6), exterior, silty clay loam; moderate coarse prismatic structure parting to weak coarse subangular blocky; extremely hard, firm; few very fine roots throughout; common very fine low continuity tubular pores; few faint discontinuous grayish brown (10YR 5/2), moist, clay films on vertical faces of peds and few prominent continuous black (N 2/0) moist, manganese or iron-manganese stains on vertical faces of peds; 10% light gray (10YR 7/1) silt deposits filling cracks; very strongly acid. (0 to 20 in thick)

GEOGRAPHIC SETTING: The Cherokee series forms in eolian sediments and alluvium in river valleys on terraces, upland benches, or in head of drains. Slopes range from 0 to 3%. Mean annual temperature ranges from 13.8 to 18.3°C (57 to 65°F). Mean annual precipitation ranges from 89 cm to 114.3 cm (35 to 45 in). The annual Thornthwaite's P.E. Index: ranges from 64 to 80. The frost-free period is 190 to 210 days.

DRAINAGE AND PERMEABILITY: The Cherokee series is somewhat poorly drained with very slow permeability and low runoff.

DISTRIBUTION AND EXTENT: Southeastern Kansas, northwestern Arkansas, southwestern Missouri, and northeastern Oklahoma. The series is moderately extensive.

SERIES ESTABLISHED: Cherokee County, Kansas, 1912.

[Adapted from National Cooperative Soil Survey Survey (2015)]

KENOMA SERIES

The Kenoma series consists of very deep, moderately well drained soils that formed in loess and/or old alluvial sediments over residuum from Pennsylvanian age shale and limestone. Kenoma soils are on plains or terraces. Slope ranges from 0 to 8%. Mean annual precipitation is 1016 mm (40 in), and mean annual temperature is 14°C (57°F).

TAXONOMIC CLASS: Fine, smectitic, thermic Vertic Argiudolls

TYPICAL PEDON: Kenoma silt loam, on an east-facing, convex, 2% slope in a native grass meadow at an elevation of 340 meters (1115 feet) above mean sea level. (Colors are for moist soil unless otherwise stated.)

A—0 to 18 cm (0 to 7 in); very dark grayish brown (10YR 3/2) silt loam, grayish brown (10YR 5/2) dry; moderate thin platy structure parting to weak fine granular; lower 8 cm (3 in) have weak fine subangular blocky structure parting to weak fine granular; slightly hard; friable; many fine roots throughout; many krotovinas; few fine subrounded chert fragments; slightly acid; abrupt wavy boundary. [13 to 33 cm (5 to 13 in) thick]

Bt1—18 to 28 cm (7 to 11 in); very dark grayish brown (10YR 3/2) silty clay, grayish brown (10YR 5/2) dry; weak fine prismatic structure parting to weak very fine subangular blocky; very firm; common fine roots throughout; few fine dendritic tubular pores; few faint very dark gray (10YR 3/1) clay films on all faces of peds; few prominent light gray (10YR 7/1) silt coats on faces of peds in upper 4 cm (1.5 in); many faint very dark brown (10YR 2/2) organic stains on vertical faces of peds; few krotovinas; common fine distinct dark yellowish brown (10YR 4/4) masses of oxidized iron in the matrix; common faint black (10YR 2/1) moderately cemented iron-manganese concretions in the matrix; few fine subrounded chert fragments; moderately acid; clear irregular boundary.

Bt2—28 to 43 cm (11 to 17 in); dark brown (10YR 3/3) silty clay, brown (10YR 4/3) dry; weak very fine subangular blocky structure; very firm; common fine roots throughout; few fine dendritic tubular pores; few distinct very dark grayish brown (10YR 3/2) clay films on all faces of peds; few faint very dark brown (10YR 2/2) organic stains on vertical faces of peds; common fine faint dark yellowish brown (10YR 4/4) masses of oxidized iron in the matrix; common faint black (10YR 2/1) moderately cemented manganese concretions in the matrix; few fine subrounded chert fragments; slightly acid; clear irregular boundary.

Bt3—43 to 66 cm (17 to 26 in); brown (10YR 4/3) silty clay, brown (10YR 5/3) dry; weak very fine subangular blocky structure; very firm; few fine roots throughout; few very fine dendritic tubular pores; few distinct dark grayish brown (10YR 4/2) clay films on all faces of peds; few faint very dark brown (10YR 2/2) organic stains on vertical faces of peds; common fine faint yellowish brown (10YR 5/4) and common fine distinct yellowish brown (10YR 5/6) masses of oxidized iron in the matrix; common fine distinct black (10YR 2/1) spherical moderately cemented manganese concretions in matrix; few fine subrounded chert fragments; neutral; gradual irregular boundary. [Combined thickness of the Bt horizon is 38 to 130 cm (15 to 51 in).]

BC—66 to 97 cm (26 to 38 in); dark yellowish brown (10YR 4/4) silty clay, yellowish brown (10YR 5/6) dry; weak very fine subangular blocky structure; firm; few fine roots throughout; common fine faint yellowish brown (10YR 5/4) masses of oxidized iron in the matrix; common fine and medium distinct black (10YR 2/1) spherical moderately cemented manganese concretions with clear boundaries in matrix; few fine subrounded chert fragments; moderately alkaline; diffuse smooth boundary. [0 to 51 cm (0 to 20 in) thick]

C1—97 to 142 cm (38 to 56 in); brown (7.5YR 4/4) silty clay, light brown (7.5YR 6/4) dry; massive; firm; common fine faint yellowish brown (10YR 5/4) masses of oxidized iron in the matrix; common fine and medium prominent black (10YR 2/1) spherical moderately cemented iron-manganese concretions with clear boundaries in the matrix; few fine subrounded chert fragments; few carbonate concretions up to 4 cm (1.5 in) in diameter; moderately alkaline; diffuse smooth boundary.

C2—142 to 152 cm (56 to 60 in); reddish brown (5YR 4/4), light brownish gray (2.5Y 6/2), pale olive (5Y 6/3), and yellowish brown (10YR 5/6) silty clay loam; massive; firm; many black (10YR 2/1) films and stains; small fragments of decomposed shale; few small carbonate concretions up to 4 cm (1.5 in) in diameter; moderately alkaline; abrupt smooth boundary.

R—152 to 200 cm (60 to 79 in), limestone bedrock.

GEOGRAPHIC SETTING: Kenoma soils are on summits, shoulders, and side slopes on interfluves of plains and also of terraces in the Cherokee Prairies. Slope ranges from 0 to 8%. The soils formed in loess and or old alluvial sediments over residuum from Pennsylvanian age shale or limestone. Mean annual precipitation ranges from 889 to 1143 mm (35 to 45 in). Mean annual temperature ranges from 13 to 18°C (55 to 64°F). Frost-free period is 175 to 225 days. Elevation is 220 to 465 meters (722 to 1525 feet) above mean sea level. Thornthwaites Annual P-E Index ranges from 60 to 80.

DISTRIBUTION AND EXTENT: MLRAs 112, 76, and 84A in southeast Kansas, southwest Missouri, and possibly northeast Oklahoma. The type location is in MLRA 112. The series is of large extent.

SERIES ESTABLISHED: Anderson County, Kansas, 1972.

[Adapted from National Cooperative Soil Survey (2014)]

Assignment: Online Quiz

A quiz will be available online. Please access it as directed by your instructor.

Subsequent Lab Set-Up

Some activities require more time than the lab period. Therefore, students must set up these activities ahead of time. The Soil Acidity and Adjusting Soil pH lab involves incubating different soils with different amounts of added lime. The goal is to study the effectiveness of various lime products on soils that have varying degrees of pH buffering capacity. We will set up the experiment in the first lab to allow soil reactions to equilibrate over the next 8 to 10 weeks. Each lab section will have six groups. Each lab group will be responsible for one of six treatments. Your laboratory instructor will assign a treatment to your group.

Sample Preparation and Set up for Adjusting Soil pH experiment

1. Label four plastic bags as follows:
    - Group name, lab section
    - Soil type
    - treatment (one of the following):
        - pure reagent grade $CaCO_3$
        - pure reagent grade CaO
        - pure reagent grade $CaSO_4$
        - dolomitic limestone (35 mesh)
        - dolomitic limestone (120 mesh)
        - control (no amendments)
2. Weigh 20 g of air-dried soil into each plastic bag.
3. Carefully weigh 0.1 g of designated liming material onto weighing paper.
4. Add liming material to soil and mix thoroughly to distribute evenly in soil.
5. Add a few mL of water to each bag and mix. Soil should be moist but not saturated. Add more water if necessary.
6. Close the bags and turn them in to your instructor.

4. Soil Classification and Mapping

The word, "taxonomy" is based on the Greek words "taxis", meaning arrangement; and "nomia", meaning method. In biology, taxonomy refers to a hierarchical system in which organisms are grouped based on shared characteristics, with domains and kingdoms at the top of the hierarchy, and genus and species at the lowest levels. Similarly, *Soil Taxonomy* is a hierarchical system used to group soils based on observable or measurable characteristics. A common application of soil classification (the act of identifying the taxonomic classification for a given soil) is to develop models of how soils of different classifications associate with one another within a landscape, which can eventually be used in soil mapping. The primary concepts of soil classification using *Soil Taxonomy* will be reviewed in this lab, followed by an overview of the Web Soil Survey (United States Department of Agriculture Natural Resources Conservation Service, 2016).

Learning Objectives

- Become familiar with the 12 soil orders.
- Understand the structure of *Soil Taxonomy*.
- Become familiar with the contents of a county soil survey report.
- Use a soil survey report for land use evaluation.

Materials

- Four soil monoliths (Crete, Clark, Morrill, and Chase soil series)
- Printed county soil survey reports
- Computer with internet access and a projector

Note

For this lab, you will need to bring a laptop or tablet to use. If you do not have a laptop or tablet, please share with a partner.

Recommended Reading

- Illustrated Guide to Soil Taxonomy (Soil Survey Staff, 2015)
- From the Surface Down: An Introduction to Soil Surveys for Agronomic Use (USDA NRCS, 2010)

Prelab Assignment

Using the assigned readings and the introduction to this lab, consider the questions listed below. These definitions/questions will provide a concise summary of the major concepts addressed in the lab. They are also relevant to the soil survey report and are useful study notes for exams.

1. Explain the difference between a profile, a pedon, and a polypedon.
2. Describe the key properties or diagnostic features for each of the following diagnostic epipedons:
    - Mollic:
    - Umbric:
    - Histic:
    - Ochric:
3. Describe the key properties or diagnostic features of the following diagnostic subsurface horizons:
    - Argillic:
    - Kandic:
    - Natric:
    - Calcic:
    - Spodic:
    - Oxic:
    - Cambic:
4. List the soil moisture regimes from dries to wettest, and note the criteria for each.
5. Explain the difference between a soil phase and a soil consociation.
6. What are the 12 soil orders? Describe the key properties or diagnostic features of each.
7. What are the six formal categories of *Soil Taxonomy*?

Introduction

Soils can vary widely in their properties, and each has a unique arrangement of layers or horizons. The *soil profile description* provides the information that distinguishes one soil from another. Review the following example of a profile description, and note the explanation of terms in Table 4.1.

Harney silt loam [adapted from the National Cooperative Soil Survey (1997)]

Ap – 0 to 9 in; dark grayish brown (10YR 4/2) silt loam, very dark grayish brown (10YR 3/2) moist; moderate medium granular structure; slightly hard, very friable; many fine roots; slightly acid; clear smooth boundary. (4 to 14 in thick)
AB – 9 to 12 in; dark grayish brown (10YR 4/2) silt loam, very dark grayish brown (10YR 3/2) moist; moderate fine subangular blocky structure; hard, friable; many fine roots; neutral; clear smooth boundary. (0 to 10 in thick)
Bt1 – 12 to 18 in; grayish brown (10YR 5/2) silty clay loam, dark grayish brown (10YR 4/2) moist; moderate medium subangular blocky structure; very hard, very firm; few fine roots; moderately alkaline; clear smooth boundary.
Bt2 – 18 to 28 in; grayish brown (10YR 5/2) silty clay loam, dark grayish brown (10YR 4/2) moist; strong medium subangular blocky structure; very hard, very firm; few fine roots; moderately alkaline; gradual smooth boundary. (Combined thickness of the Bt horizon is 10 to 26 in)
BCk2 – 8 to 35 in; brown (10YR 5/3) silty clay loam, brown (10YR 4/3) moist; moderate medium subangular blocky structure; hard, firm; few fine roots; many soft accumulations of carbonates; strong effervescence; moderately alkaline; gradual smooth boundary. (0 to 16 in thick)
Ck3 – 5 to 47 in; pale brown (10YR 6/3) silt loam, brown (10YR 5/3) moist; massive; slightly hard, friable; common soft accumulations of carbonates; strong effervescence; moderately alkaline; gradual smooth boundary. (0 to 20 in thick)
C – 47 to 60 in; pale brown (10YR 6/3) silt loam, brown (10YR 5/3) moist; massive; slightly hard, friable; strong effervescence; moderately alkaline.

Table 4.1. Explanation of the Ap horizon description.

Morphological property	Description	Morphological property	Description
Horizon designation	Ap	Structure grade	Moderate
Upper depth	0 in	Structure size	Medium
Lower depth	9 in	Structure type	Granular
Color name	Dark grayish brown	Moist consistence	Very friable
Munsell hue	10YR	Roots	Many fine roots
Munsell value	4	pH	Slightly acid
Munsell chroma	2	Boundary	Clear smooth
Textural class	Silt loam		

Table adapted from King et al. (2003).

Completing a soil profile description involves a systematic approach:

1. Observing the landscape setting.

2. Examining the morphological features like texture, structure, color, consistence, etc. of the soil to distinguish any layers or horizons
3. Describing in detail the texture, structure, color, consistence, and other important features of each horizon.
4. Assigning horizon designations to each layer.
5. Classifying the soil on the basis of its morphology and horizonation.

Soil Morphology and Land Use

Criteria that rate soils for a particular use are important to land use planning and land management decisions. Guidelines based on these criteria facilitate uniform and consistent land evaluations. Soil-based criteria can be developed for nearly any land use. To prepare a soil rating scheme, the following are required:

- Precise definition of the land use
- A list of soil properties affecting the use
- Limits for each soil property that would be favorable or unfavorable for the land use.

A comprehensive classification system is important for any science: soil science, plant science, biology, geology, among many others. Effective taxonomy allows us to organize knowledge and learn new relationships. *Soil Taxonomy* helps in extrapolating soil management research among similar soils around the world. *Soil Taxonomy* is a quantitative system based on soil properties that can be observed or measured, organized in a hierarchy based on six categories beginning with 12 broad soil orders and narrowing in specificity to more than 23,000 series. The following diagram illustrates the organization of a taxonomic name by category.

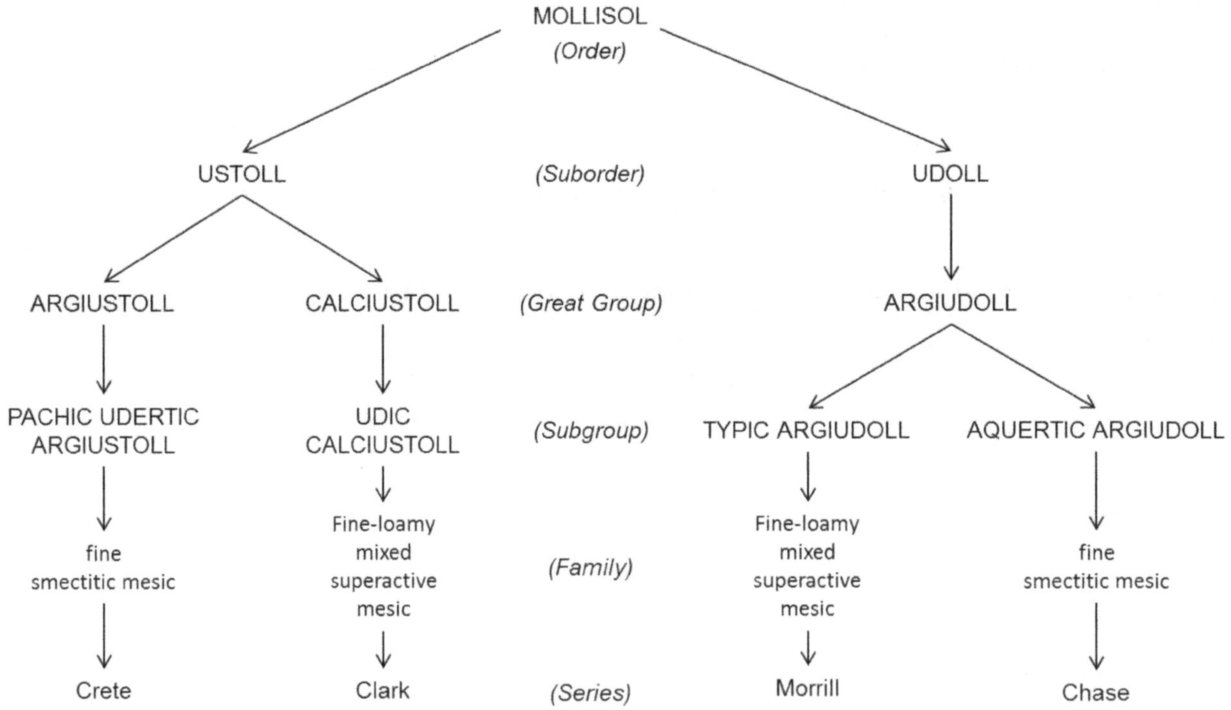

Figure 4.1. An example of USDA Soil Taxonomy classifications for four soil series within the Mollisol soil order.

Table 4.2 Simplified key to the 12 soil orders

Order	Major Diagnostic Features	Formative Element
Gelisols	Soils with permafrost or gelic material within 100 cm	el
Histosols	Other soils with >30% organic matter (>12% organic carbon) content to a depth of 40 cm or more	ist
Spodosols	Other soils with a spodic horizon (illuvial humus, iron) within a depth of 200 cm	od
Andisols	Other soils with andic soil properties (low density, volcanic glass, pumice, etc.) in >= 50% of the upper 60 cm	and
Oxisols	Other soils with an oxic horizon, or containing more than 40% clay in the surface 18 cm and a kandic horizon with less than 10% weatherable minerals (highly weathered)	ox
Vertisols	Other soils containing more than 30% clay in all horizons and cracks that open and close periodically (shrinking/swelling)	ert
Aridisols	Other soils with some diagnostic subsoil horizon(s) and an aridic soil moisture regime	id
Ultisols	Other soils with an argillic or kandic horizon and a base saturation at pH 8.2 of <35% at a depth of 180 cm	ult
Mollisols	Other soils with a Mollic epipedon and a base saturation at pH 7 of ≥50% in all depths above 180 cm	oll
Alfisols	Other soils with an argillic, kandic, or natric horizon (and a base saturation at pH 8.2 of >35% at a depth of 180 cm)	alf
Inceptisols	Other soils with an umbric, mollic, or plaggen epipedon or a cambic horizon	ept
Entisols	Other soils	ent

The formative elements are used in the names of suborders and lower taxonomic levels. Table courtesy of R. Weil.

Many other formative elements can specify unique soil properties at each taxonomic level. Each formative element has a connotation for a given soil. These connotations of the formative elements used for suborders and great groups are listed in Table 4.3 and Table 4.4.

Table 4.3. Formative elements used to identify various suborders in Soil Taxonomy.

Formative Element	Connotation	Formative Element	Connotation
alb	Presence of albic horizon (a bleached eluvial horizon)	hist	Presence of histic epipedon
anthr	Presence of anthropic or plaggen epipedon	hum	Presence of organic matter
aqu	Characteristics associated with wetness	orth	The common ones
ar	Mixed horizons	per	Of year-round humid climates, perudic moisture regime
arg	Presence of argillic horizon (a horizon with illuvial clay)	psamm	Sand textures
calc	Presence of calcic horizon	rend	Rendzinalike-high in carbonates
camb	Presence of cambric horizon	sal	Presence of salic (saline) horizon
cry	cold	sapr	Most decomposed stage
dur	Presence of a duripan	torr	Usually dry
fibr	Least decomposed stage	turb	Cryoturbation
fluv	Floodplains	ud	Of humid climates
fol	Mass of leaves	ust	Of dry climates, usually hot in summer
gyps	Presence of gypsic horizon	vitr	Resembling glass
hem	Intermediate stage of decomposition	xer	Dry summers, moist winters

Table from King et al. (2003).

Table 4.4 Formative elements for names of great groups and their connotations

Formative Element	Connotation	Formative Element	Connotation
acr	Extreme weathering	hist	Presence of organic materials
aer	Chroma >2, non-reducing	fragi	Fragipan
agr	Agric horizon	hum	Humus
al	High aluminum, low iron	hydr	Water
alb	Albic horizon	kand	Low activity 1:1 silicate clay
and	Ando-like	lithic	Near stone
anhy	Anhydrous	luv, lu	Illuvial
aqu	Water saturated	melan	Melanic epipedon
aren	Sandy	molli	With a mollic epipedon
argi	Argillic horizon	natr	Presence of a natric horizon
calc, calci	Calcic horizon	pale	Old development
camb	Cambric horizon	petr	Cemented horizon
chrom	High chroma	plac	Thin pan
cry	Cold	plagg	Plaggen horizon
dur	Duripan	plinth	Plinthite
dystr, dys	Low base saturation	psamm	Sand texture
endo	Fully water saturated	quartz, quartzi	High quartz
epi	Perched water table	rhod	Dark red colors
eutr	High base saturation	sal	Salic horizon
ferr	Iron	sapr	Most decomposed
fibr	Least decomposed	somb	Dark horizon
fluv	Floodplain	sphagn	Sphagnum moss
fol	Mass of leaves	sulf	Sulfur
fragloss	See frag and gloss	torr	Usually dry and hot
fulv	light-colored melanic horizon	ud	Humid climates
gyps	gypsic horizon	umber	Umbric epipedon
gloss	Tongued	ust	Dry climate, usually hot in summer
hal	Salty	verm	Wormy or mixed by animals
hapl	Minimum horizon	vitr	Glass
hem	Intermediate decomposition	xer	Dry summers, moist winters

Table adapted from King et al. (2003).

A complete taxonomic name communicates a great deal of information about the soil if we understand each part of the name. As an example of the quantitative information revealed in a taxonomic name, the following classification name will be dissected by category. Consider, for example, the Harney soil, with a taxonomic classification of fine, smectitic, mesic Typic Argiustoll.

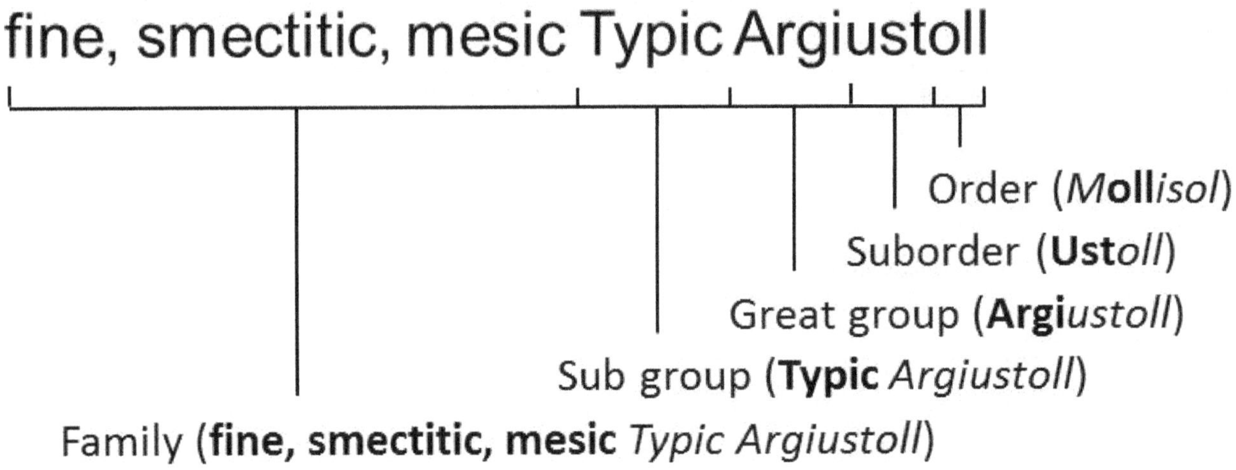

Figure 4.2. Formative elements in the taxonomic classification of the Harney series.

Table 4.5.Translation of the taxonomic classification of the Harney Series.

Categories	Properties connoted
ORDER: Mollisol	Has a mollic epipedon and a base saturation of >50% to a depth of 1.8 m from the soil surface or to an impermeable layer
SUBORDER: Ustoll	has an ustic moisture regime; dry for as long as 90 days cumulatively per year
GREAT GROUP: Argiustoll	has an argillic horizon
SUBGROUP: Typic Argiustoll	typical of an Argiustoll, not intergrading toward another great group condition
FAMILY: fine, smectitic, mesic	the upper 50 cm of the argillic horizon has 35-60% clay; the dominant clay minerals are smectite minerals (montmorillonite, beidellite, and nontronite); the mean annual soil temperature at 50 cm is 8°C to 15°C (47°F to 59°F)
SERIES: Harney	differs from soils in the same family in based on color, parent material (loess), and calcium accumulation below 28 in.

Table courtesy of C. J. Moorberg, adapted from King et al. (2003).

Activity 1: Practice Key to Soil Orders

Now that you have studied the characteristics of the 12 soil orders, enter the most appropriate soil order name in each rectangle.

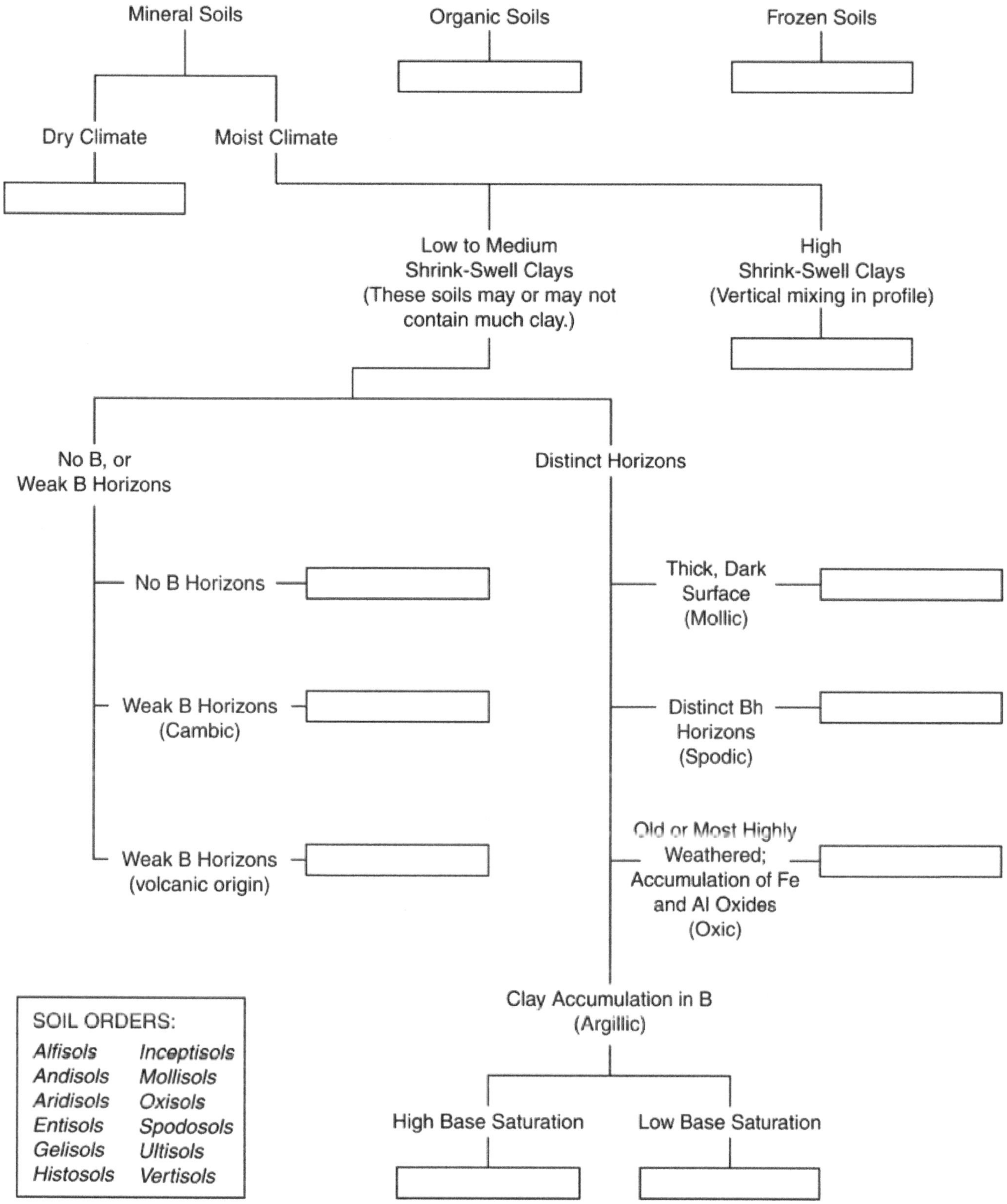

Figure 4.3. General characteristics of the 12 soil orders.

Activity 2: Structure of *Soil Taxonomy*

To illustrate the structure of *Soil Taxonomy*, separate a complete taxonomic name into the six categories. Follow the example of the Harney silt loam in figure 4.2.

Colby

Taxonomic Name: Fine-silty, mixed, superactive, calcareous, mesic Aridic Ustorthents

Order
Suborder
Great Group
Subgroup
Family
Series

Goessel

Taxonomic Name: Fine, smectitic, mesic Typic Haplusterts

Order
Suborder
Great Group
Subgroup
Family
Series

Wymore

Taxonomic Name: Fine, smectitic, mesic Aquertic Argiudolls

Order
Suborder
Great Group
Subgroup
Family
Series

Activity 3: Interpreting Taxonomy

As a further exercise in understanding taxonomic names, complete the following questions. Use the list of taxonomic names of soils representative of Mollisols from the prairie pothole region of Iowa below to answer these questions.

The Des Moines lobe of the Wisconsin glaciation covered north-central Iowa with a deep layer of glacial deposits, and provides a good example of how taxonomic names depict important soil properties. The Clarion-Nicollet-Webster-Glenco topo-sequence, or "catena" (Figure 4.4), illustrates how *Soil Taxonomy* reflects wetness, or depth to a water table.

Fill in the subgroup taxonomic name for soils in Table 4.6, and study how the terms change with wetness.

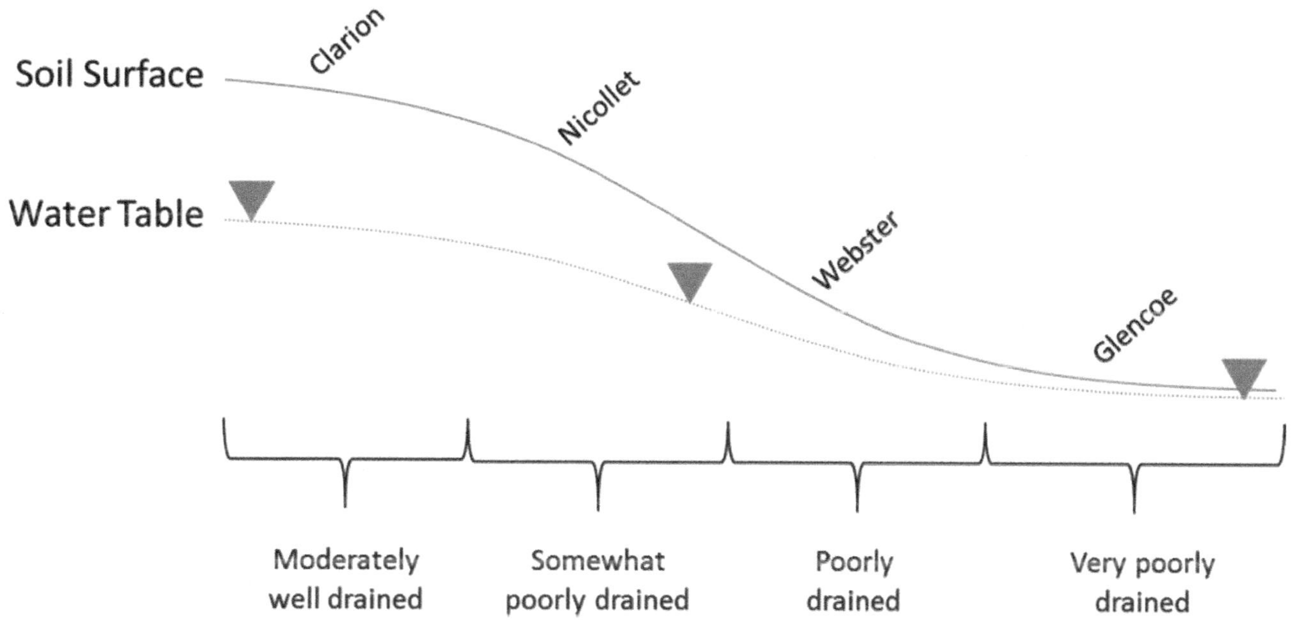

Figure 4.4. A cross-section depicting the Clarion-Nicollet-Webster-Glencoe catena. Figure courtesy of C. J. Moorberg.

Clarion series: Fine-loamy, mixed, superactive, mesic Typic Hapludolls

Nicollet series: Fine-loamy, mixed, superactive, mesic Aquic Hapludolls

Webster series: Fine-loamy, mixed, superactive, mesic Typic Endoaquolls

Glencoe series: Fine-loamy, mixed, superactive, mesic Cumulic Endoaquolls

Table 4.6. Clarion-Nicollet-Webster-Glenco topo-sequence Taxonomy.

Series	Drainage Class	Depth to Seasonal High Water Table	Subgroup Taxonomic Classification
Clarion	Moderately Well	61 - 102 cm (24 - 48 in)	
Nicollet	Somewhat Poorly	30 – 61 cm (12 – 24 in)	
Webster	Poorly	< 30 cm (< 12 in)	
Glencoe	Very Poorly	< 30 cm (< 12 in), and accumulation of organic matter	

Table courtesy of C. J. Moorberg.

Notice that the wetter the drainage class (that is, the shallower the depth to the seasonal high water table), the higher the "aqu" formative element becomes in the overall classification. That is because *Soil Taxonomy* prioritizes soil management considerations. The depth to the seasonal high water table would be a management concern for most land uses for the Nicollet, Webster, and Glencoe series; it would be of less concern for the Clarion series, and thus "aqu" is not included in the classification.

Also note that for the Glencoe series, in addition to having the "aqu" formative element as part of the suborder, the "cumulic" formative element has been designated in the subgroup. That formative element alludes to a "thickened epipedon" caused by the accumulation of organic matter. Because the water table is so shallow, little oxygen is available at the surface for a significant portion of the growing season. This slows decomposition, allowing organic matter to build, thus creating a thickened epipedon with lots of organic matter.

Activity 4: Practicing *Soil Taxonomy* Interpretations with State Soils of the US

State soils have been selected for all 50 states and three territories in the U.S. The group of soils represents a diverse sample of soil conditions and classifications. It serves as an interesting focus for a little practice at deciphering and understanding *Soil Taxonomy*. Use the attached list of state soils in Table 4.7 along with Table 4.2, Table 4.3, and Table 4.4 to answer the following questions:

- What is the most commonly recognized ORDER among the state soils?
- Which of the soil ORDERS is not represented in the list of state soils?
- How many Oxisols are represented by the 53 soils?
- What is the complete SUBORDER name for the state soil of Alaska?
- How many Vertisols are represented in the state soils?

In what soil property does the Downer soils of New Jersey differ from the Greenwich soils or Deleware?

The state soil of South Carolina has a soil condition identified by its great group. What element is present in the upper 50 cm of this soil? (Hint: use Table 4.4)

What is the complete taxanomic name for the state soil of Kansas?

Table 4.7. Soil Taxonomy classifications of state soils of the U.S.

Series	State	Family Classification
Tanana	AK	coarse-loamy, mixed, superactive, subgelic Typic Aquiturbels
Bama	AL	fine-loamy, siliceous, subactive, thermic Typic Paleudults
Stuttgart	AR	fine, smectitic, thermic Albaquultic Hapludalfs
Casa Grande	AZ	fine-loamy, mixed, superactive, hyperthermic Typic Natrargids
San Joaquin	CA	fine, mixed, active, thermic Abruptic Durixeralfs
Seitz	CO	clayey-skeletal, smectitic Ustic Glossocryalfs
Windsor	CT	mixed, mesic Typic Udipsamments
Greenwich	DE	coarse-loamy, mixed, semiactive, mesic Typic Hapludults
Myakka	FL	sandy, siliceous, hyperthermic Aeric Alaquods
Tifton	GA	fine-loamy, kaolinitic, thermic Plinthic Kandiudults
Akina	GU	very-fine, kaolinitic, isohyperthermic Inceptic Haplustox
Hilo	HI	medial over hydrous, ferrihydritic, isohyperthermic Acrudoxic Hydrudands
Tama	IA	fine-silty, mixed, superactive, mesic Typic Argiudolls
Threebear	ID	medial over loamy, amorphic over mixed, superactive, frigid Oxyaquic Udivitrands
Drummer	IL	fine-silty, mixed, superactive, mesic Typic Endoaquolls
Miami	IN	fine-loamy, mixed, active, mesic Oxyaquic Hapludalfs
Harney	KS	fine, smectitic, mesic Typic Argiustolls
Crider	KY	fine-silty, mixed, active, mesic Typic Paleudalfs
Ruston	LA	fine-loamy, siliceous, semiactive, thermic Typic Paleudults
Paxton	MA	coarse-loamy, mixed, active, mesic Oxyaquic Dystrudepts
Sassafras	MD	fine-loamy, siliceous, semiactive, mesic Typic Hapludults
Chesuncook	ME	coarse-loamy, isotic, frigid Aquic Haplorthods
Kalkaska	MI	sandy, isotic, frigid Typic Haplorthods
Lester	MN	fine-loamy, mixed, superactive, mesic Mollic Hapludalfs
Menfro	MO	fine-silty, mixed, superactive, mesic Typic Hapludalfs
Natchez	MS	coarse-silty, mixed, superactive, thermic Typic Eutrudepts
Scobey	MT	fine, smectitic, frigid Aridic Argiustolls
Cecil	NC	fine, kaolinitic, thermic Typic Kanhapludults
Williams	ND	fine-loamy, mixed, superactive, frigid Typic Argiustolls
Holdrege	NE	fine-silty, mixed, superactive, mesic Typic Argiustolls
Marlow	NH	coarse-loamy, isotic, frigid Oxyaquic Haplorthods
Downer	NJ	coarse-loamy, siliceous, semiactive, mesic Typic Hapludults
Panistaja	NM	fine-loamy, mixed, superactive, mesic Ustic Haplargids
Orovada	NV	coarse-loamy, mixed, superactive, mesic Durinodic Xeric Haplocambids
Honeoye	NY	fine-loamy, mixed, semiactive, mesic Glossic Hapludalfs
Miamian	OH	fine, mixed, active, mesic Oxyaquic Hapludalfs
Port	OR	fine-silty, mixed, superactive, thermic Cumulic Haplustolls
Hazleton	PA	loamy-skeletal, siliceous, active, mesic Typic Dystrudepts
Bayamon	PR	very-fine, kaolinitic, isohyperthermic Typic Hapludox

Series	State	Family Classification
Narragansett	RI	coarse-loamy over sandy or sandy-skeletal, mixed, active, mesic Typic Dystrudepts
Bohicket	SC	fine, mixed, superactive, nonacid, thermic Typic Sulfaquents
Houdek	SD	fine-loamy, mixed, superactive, mesic Typic Argiustolls
Dickson	TN	fine-silty, siliceous, semiactive, thermic Glossic Fragiudults
Houston Black	TX	fine, smectitic, thermic Udic Haplusterts
Taylorsflat	UT	fine-loamy, mixed, superactive, mesic Xeric Haplocalcids
Pamunkey	VA	fine-loamy, mixed, semiactive, thermic Ultic Hapludalfs
Victory	VI	loamy-skeletal, mixed, superactive, isohyperthermic Typic Haplustepts
Tunbridge	VT	coarse-loamy, isotic, frigid Typic Haplorthods
Tokul	WA	medial, amorphic, mesic Aquic Vitrixerands
Antigo	WI	coarse-loamy over sandy or sandy-skeletal, mixed, superactive, frigid Haplic Glossudalfs
Monongahela	WV	fine-loamy, mixed, semiactive, mesic Typic Fragiudults
Forkwood	WY	fine-loamy, mixed, superactive, mesic Ustic Haplargids

Table courtesy of J. Kleiss and D. Lindbo.

Activity 5: Soil Survey Reports

As an introduction to soil reports, look through a typical printed county soil survey report; take note of the manual's organization and the extensive content. The report begins with some background information on the county, along with an overview of how the survey was conducted. The county soil conditions are described, and the soil mapping units are discussed in detail. A colored map displays these general soil units.

Following the brief overview are detailed soil map unit descriptions. These show the symbol that is on the soil map, the dominant soil series, topsoil texture and range of slope found in the unit. The descriptions of each map unit details the landscape setting, general properties, and major use and management considerations. If other soils are present in the map unit, this is an important part of map unit description. The next section of the soil report offers specific use and management suggestions and discusses how specific types of land use ratings were formulated. This is followed with an overview of what specific kinds of soil data are included.

Specific information on soil classification and detailed profile descriptions for each soil are followed by a glossary of terms used in the report. A sequence of tables provides detailed ratings on a wide range of land uses. This interpretative information is offered for each soil map unit. The last section of the report shows the soil maps on an aerial photograph base.

To become familiar with county soil survey reports select one provided and review the table of contents and the summary list of tables. Leaf through the report and note the following sections:

- Map Unit descriptions
- Use and management of soils
- Classification and profile descriptions
- Interpretive tables

- General soil map
- Soil legend
- Soil maps

The United States Department of Agriculture Natural Resource Conservation Service (USDA NRCS) today provides these soil surveys in a digital format through the Web Soil Survey (United States Department of Agriculture Natural Resources Conservation Service, 2016). The Web Soil Survey provides all the information previously contained in the county soil survey reports. It also contains additional tools and information that has not been available in printed versions of the soil surveys. Another advantage of the Web Soil Survey is that the information contained in it can be updated as needed, instead of being updated following new surveys of the same county, which take 30 to 60 years! Your instructor will walk you through some of the main features of the Web Soil Survey and show you how to request a PDF copy of a soil survey report for a designated area. You will use these skills for your Soil Survey Report assignment.

Assignment: Soil Survey Report

For this lab, you will be preparing a soil survey report. The report assignment will be provided to you at the beginning of the lab. Your instructor will go over what to include in the report and where to collect the necessary information from the Web Soil Survey.

Subsequent Lab Set-Up

Some activities require preparation beyond the lab period and must be set up ahead of time. The soil texture by hydrometer activity in the Soil Texture and Structure lab involves dispersing soil particles chemically, which requires time for the reactions to take place. We will do this now, so the samples are ready next week.

For each of the three soils provided, do the following:

Weigh out 30.0 g of dry soil (assume oven-dry) into a 250-ml Erlenmeyer flask.

Wash sides of flask with distilled water from a wash bottle.

Add 100 ml of distilled water using a graduated cylinder. Then add 10 ml of sodium hexametaphosphate solution (500 g/L) from the dispenser on the sodium hexametaphosphate bottle.

Swirl to mix.

Wash sides of flask with distilled water from a wash bottle.

Cover the flask with Parafilm and label the flask with your lab section, table number, and soil type. Store the flasks in the location specified by your instructor for the next laboratory period.

5. SoilWeb Field Trip

Soil information is critical to managing our landscape. There were soil mapping efforts in the US prior to the Dust Bowl. However, the need to implement conservation practices on farmland following the Dust Bowl ultimately led to the creation of what's now known as the US Department of Agriculture Natural Resources Conservation Service (USDA-NRCS) as well as that agency's primary missions – mapping the soils of the US to support conservation efforts. The importance of the nation-wide soils map data has expanded beyond implementing conservation practices on farmland, and is now used by many different disciplines. Examples include real estate development, habitat restoration, land valuation, and more. The USDA-NRCS leads soil mapping efforts in cooperation with many other federal, state, and local government agencies and Land-Grant institutions from around the country as part of the National Cooperative Soil Survey (NCSS). The resulting soils data is contained in five soil databases. Most of the time we access soil data from the SSURGO database, which is the same database used for the Web Soil Survey. Other services and applications also use this database, including the SoilWeb web application that we will use in this self-led laboratory.

Learning Objectives

- Explore soils in your local area.
- Demonstrate proficiency in using the SoilWeb web application.
- Become familiar with data available in the SSURGO database.

Materials

- Using the SoilWeb Web Application Problem Set
- A smartphone or tablet with access to the internet
- SoilWeb web application

Recommended Reading & Viewing

- SoilWeb Apps (O'Geen, 2021)
- Description of SSURGO Database (USDA-NRCS Soil Survey Staff, 2021)
- From the Surface Down: An Introduction to Soil Surveys for Agronomic Use (USDA NRCS, 2010)

Prelab Assignment

Using the recommended reading resources, consider the questions listed below.

1. What is the difference between a "web application" (or "web app" for short) and a native application (app) installed on a phone or tablet?
2. List and briefly describe each of the other applications available that are associated with SoilWeb.
3. Based on the "Status Map of Soil Surveys Available from the Web Soil Survey" on the "Description of SSURGO Database" web page, what state has the most area remaining that needs to be mapped?
4. What is the range of scale at which soils are mapped for the SSURGO database, and what is the most common scale?

Appropriate Clothing

We recommend wearing long pants, sturdy shoes or boots, and socks that extend past your ankles on this field trip. Sunscreen and bug spray may also be needed.

Physical Ability

This is a self-led field trip, and you will decide what area you will explore. We recommend you choose a safe area that you can walk or hike within your abilities, or that you find an alternative means of transportation as needed.

Assignment: Problem Set

A problem set will be provided to you. The questions in the problem set will guide you through the laboratory activity.

6. Soil Pit Field Trips

One of the fundamental skills of a soil scientist is to accurately describe a soil profile. To do so, soil scientists need to have a thorough understanding of many concepts that have already been covered in this course, including soil color, texture, structure, taxonomic classification, soil development processes, and the five soil forming factors. Using those soil science concepts, you will be working with your classmates in groups to fill out a soil profile description sheet at two different sites over the course of two field trips. One site has been used for row crop agriculture, while the other is managed range land. Through these two land uses we can compare how two different management systems impacted these two particular soil profiles. We will also discuss the soil mapping process, and appropriate interpretation and use of data from the Web Soil Survey.

Learning Objectives

- Describe a soil profile using a soil profile description sheet.
- Delineate soil profile horizons.
- Interpret a soil profile using the five soil forming factors.
- Identify the impacts of management on soil properties.

Materials

- Cultivated and Prairie Soil Comparison Problem Set
- Trowels
- Shovels
- Muffin tins
- Munsell Soil Color Charts
- Clinometers
- Water squirt bottles
- Hydrochloric acid, 10%
- Soil profile measuring tape and a nail
- Profile horizon markers

Recommended Reading & Viewing

- Eye on Agriculture Today: Soil Texture by Feel (KSREVideos, 2010)
- Soil Profiling: Color (KSREVideos, 2011a)
- Soil Profiling: Structure (KSREVideos, 2011b)
- Soil Profiling: The Proper Tools (KSREVideos, 2011c)
- Illustrated Guide to Soil Taxonomy (Soil Survey Staff, 2015)

Prelab Assignment

Using the recommended reading and viewing resources and the materials for the Soil Formation, Soil Classification and Mapping, and Soil Texture and Structure labs, consider the questions listed below.

1. Define epipedon, and name one example.
2. Define subsurface diagnostic horizon, and name one example.
3. What is the difference between consociations, complexes, and associations?
4. List the landscape positions on a hillslope from top to bottom. Which position is the most stable?
5. When labeling soil profile horizons, how does a number to the left of the master horizon (example: 2Bt) differ from a number to the right of the master horizon (example: Bt2)?
6. In a few words for each, identify the soil profile feature indicated by each subordinate horizons.
    1. g:
    2. k:
    3. n:
    4. p:
    5. t:
    6. ss:

Appropriate Clothing

We recommend wearing long pants, sturdy shoes or boots, and socks that extend past your ankles on this field trip. Sunscreen and bug spray may also be needed.

Physical Ability

This field trip may require 5 to 10 minute hikes over uneven terrain. It may also entail climbing eight foot tall fences. If you cannot perform these tasks, contact your instructor as soon as possible. An alternative laboratory and/or field trip activity will be organized for you.

Assignment: Problem Set

A problem set will be handed out at the start of this lab. The questions in the problem set focus the soil pit field trip(s).

Subsequent Lab Setup

To examine the influence of texture and density on water holding capacity and pore size distribution, the moisture status of different soil materials will be monitored through a drying cycle. Each lab group will be responsible for two soils and one treatment (Air Dry, AD; Wilting Point, WP; Field Capacity, FC; and Saturated, S). Your laboratory instructor will assign the treatment. You will complete the experiment on both soils.

- Label an empty cup as follows:
    - Lab Section
    - Soil Type: Sandy or Clayey
    - Treatment:AD (air dry)
    - WP (wilting point)
    - FC (field capacity)
    - S (saturated)
- Punch eight small holes in the bottom of the cup.
- Weigh the empty cup and record the value on the data sheet.
- Cut a small piece of paper towel and place in the bottom of the cup to prevent soil from falling through the holes.
- Place a cup on the balance, tare out the weight of the cup, and add exactly 50g of soil and gently tap the cup on the lab bench to settle the soil.
- We need to know the volume of soil in the cups so we can calculate bulk density. Fill another cup (without holes!) with water to the same level as the soil in the cups. Pour the water into a graduated cylinder and record the value on the data sheet. (assume 1 ml = 1 cm^3)
- Get the data for the other soil from the lab groups with the other soil.
- Calculate and record the bulk density.

- Carefully and slowly place the WP, FC, and S cups in the water in the dishpans provided so the soil can become saturated from the bottom to the top. Your lab instructor will remove WP cups tomorrow and allow them to dry for 6 days. The FC cups will be removed in 6 days and allowed to drain for 1 day. The S cups will remain in the water all week.
- Place the AD cup in the dishpan without water.

PART III

SOIL PHYSICS

7. Soil Texture and Structure

Soil texture and structure are considered "master variables", meaning that texture and structure directly influence a large number of other soil properties. For example, in comparing a clayey soil and a sandy soil, one would expect the clayey soil to have larger specific surface area, more cation exchange capacity, more total porosity, less macroporosity, and more organic matter than the sandy soil. Thus, by simply knowing the texture of the soil, inferences can be made in regard to many soil properties. Here, soil texture will be determined quantitatively using the hydrometer method, and estimated using the texture by feel method. Different soil structure types will also be observed. What you learn about texture and structure in these activities will be used later during the soil pit field trips to describe soil profiles in the field.

Learning Objectives

- Differentiate the three soil separates (sand, silt, and clay) based on their particle size diameters.
- Determine the percentages of sand, silt, and clay in selected soil samples using data collected from the hydrometer method of particle size analysis.
- Estimate the textural class using the texture-by-feel method on selected soil samples.
- Use a textural triangle to determine the textural class of a soil.
- Understand the relationship between particle size and specific surface area.

Materials

- Three soils of known textures
- Sodium hexametaphosphate
- Squirt bottles filled with tap water
- Lab balances accurate to the nearest 0.01 g
- Soil test cylinders, 1L volume (Item #200231000, Kimble ™ Kimax™ Soil Test Cylinders)
- Hydrometers calibrated in units of g/L (Item #13-202-133 Fisherbrand™ Soil Analysis ASTM Hydrometer)
- Digital thermometers accurate to the nearest 0.1°C
- Milkshake mixers and stainless steel milkshake mixer cups (Hamilton Beach HMD400 120V Triple Spindle Commercial Drink Mixer, Hamilton Beach, Glen Allen, Virginia, U.S.)

Recommended Reading

- Eye on Agriculture Today: Soil Texture by Feel (KSREVideos, 2010)
- Soil Profiling: Structure (KSREVideos, 2011c)
- Estimating Soil Texture by Feel (Presley and Thien, 2008)

Prelab Assignment

Using the recommended reading and viewing resources and the introduction to this lab, consider the questions listed below. These definitions/questions will provide a concise summary of the major concepts to be addressed in the lab. They will also serve as the basis for the post-lab quiz and are useful study notes for exams.

1. Define and explain the difference between soil texture and soil structure.
2. List the sizes for sand, silt, and clay particles using USDA criteria.
3. What is a textural triangle?
4. What is Stoke's Law and how is it used in soil science?
5. Define specific surface area and explain its relationship to size of soil particles.

Introduction

Soil texture refers to the proportions of sand (2.0 - 0.05 mm in diameter), silt (0.05 - 0.002 mm), and clay (less than 0.002 mm). The relative proportions determine the textural class. Soil texture influences nearly every aspect of soil use and management. Many of the physical and chemical properties of the soil depend on how fine (clayey) or coarse (sandy) a soil is. Soil texture is a permanent feature unless soils are subjected to rapid erosion, deposition, or removal.

Moreover, much of the reactivity of soils is related to the amount of surface area available. As the average particle size decreases, the surface area per unit weight increases (see Table 7.1).

Table 7.1. Surface areas of soil particle sizes.

Seperate	Diameter (mm)	Particles/g	Specific surface area (cm^2/g)
Very course sand	2.00 - 1.00	90	11
Coarse sand	1.00 - 0.50	720	23
Medium sand	0.50-0.25	5,700	45
Fine sand	0.25-0.10	46,000	91
Very fine sand	0.10-0.05	722,000	277
Silt	0.05-0.002	5,776,000	454
Clay	< 0.002	90,260,000,000	8,000,000

Table from King et al. (2003).

Nearly any type of land management will be influenced by texture. Table 7.2 provides a summary of soil management factors related to texture.

Table 7.2. Summary of soil texture relationships to various soil physical and chemical properties.

Water relations	Sandy soils	Loamy soils	Clayey soils
Infiltration—entry of surface water into the soil. Opposite of run-off potential	Rapid	Medium to slow	Very rapid if cracks present; slow if no cracks
Percolation—internal water drainage and leaching	Excessive	Good	Fair-Poor
Water Storage—available for plant use	Very low	Medium	High
Aeration—movement of oxygen into root zone	Very good	Moderate	Poor
Tillage and Erosion Tillage power required	Low	Medium	High
Tillability—ease of seed-bed preparation	Easy	Medium	Difficult
Erodibility Wind erosion hazard	High	Low	Medium
Water erosion hazard	Low	High	Low to medium
Chemical Relations Fertility potential (nutrient storage)	Low	Medium	High
Chemical Recommendations—rates per acre	Low	Medium	High

Table from King et al. (2003).

Activity 1: Textural Triangle

Soils with similar distributions of sand, silt, and clay have similar properties and are therefore grouped into the same soil textural class. Twelve textural classes are recognized, and their compositions are designated on a textural triangle (Figure 7.1). Study the arrangement of the triangle. Each corner represents 100% of either sand, silt, or clay, and each represents 0 to 100% of a given fraction. The proportions of sand, silt, and clay define the twelve classes. Figure 7.1 depicts a soil with 20% sand, 25% silt, and 55% clay. Those three lines intersect within the boundaries of the "Clay" textural class, so the soil is a clay soil.

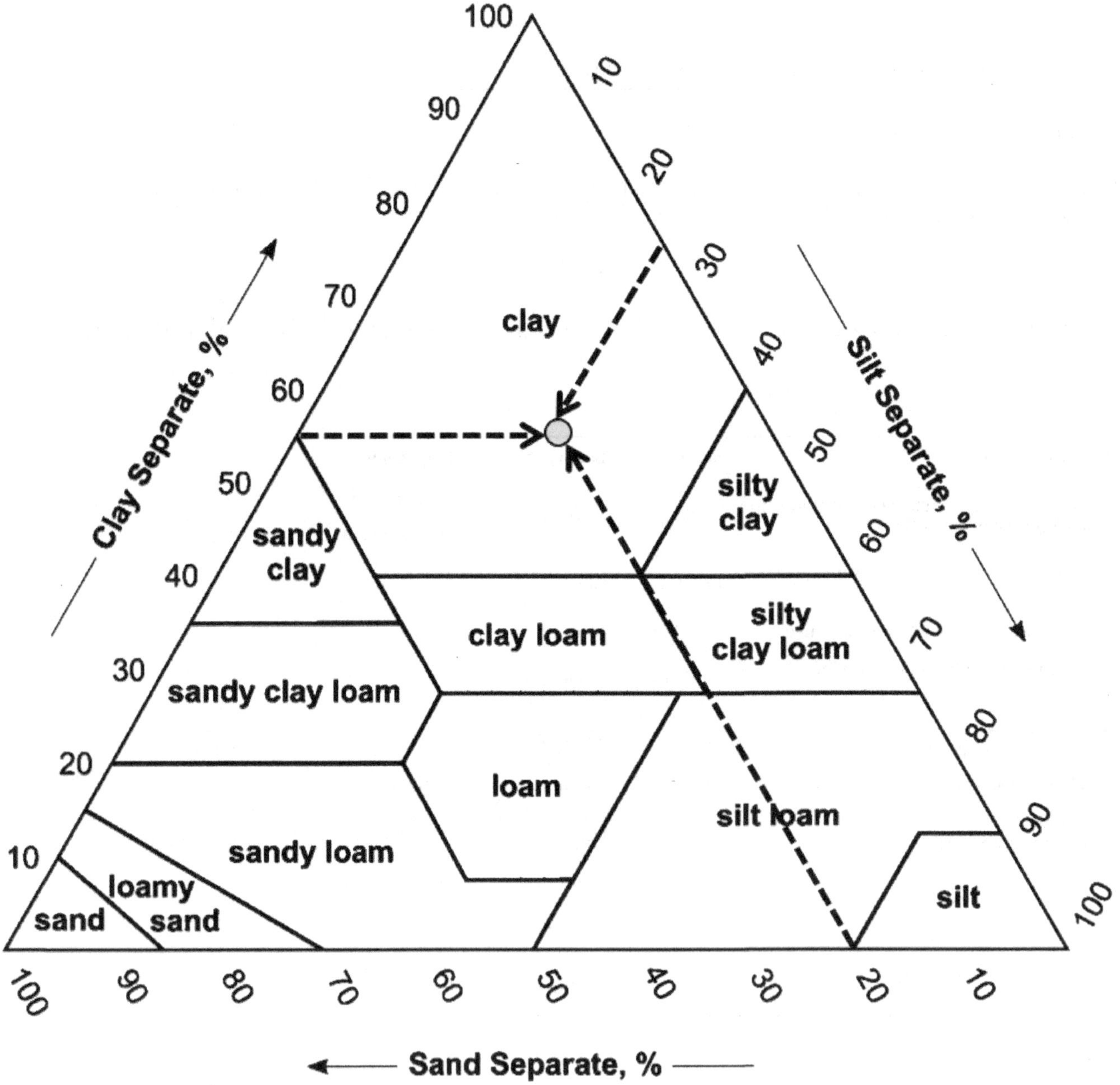

Figure 7.1. Soil texture triangle example. Figure adapted from King et al. (2003).

Using that same procedure, determine the textural class names for the soils described in Table 7.3, and plot each result on the textural triangle in Figure 7.2.

Table 7.3. Soil Texture Class Activity

	% Sand	% Silt	% Clay	Textural Class
1.	33	33	34	
2.	55	30	15	
3.	80	5	15	
4.	25	60	15	
5.	-	20	50	
6.	60	-	30	
7.	40	40	-	

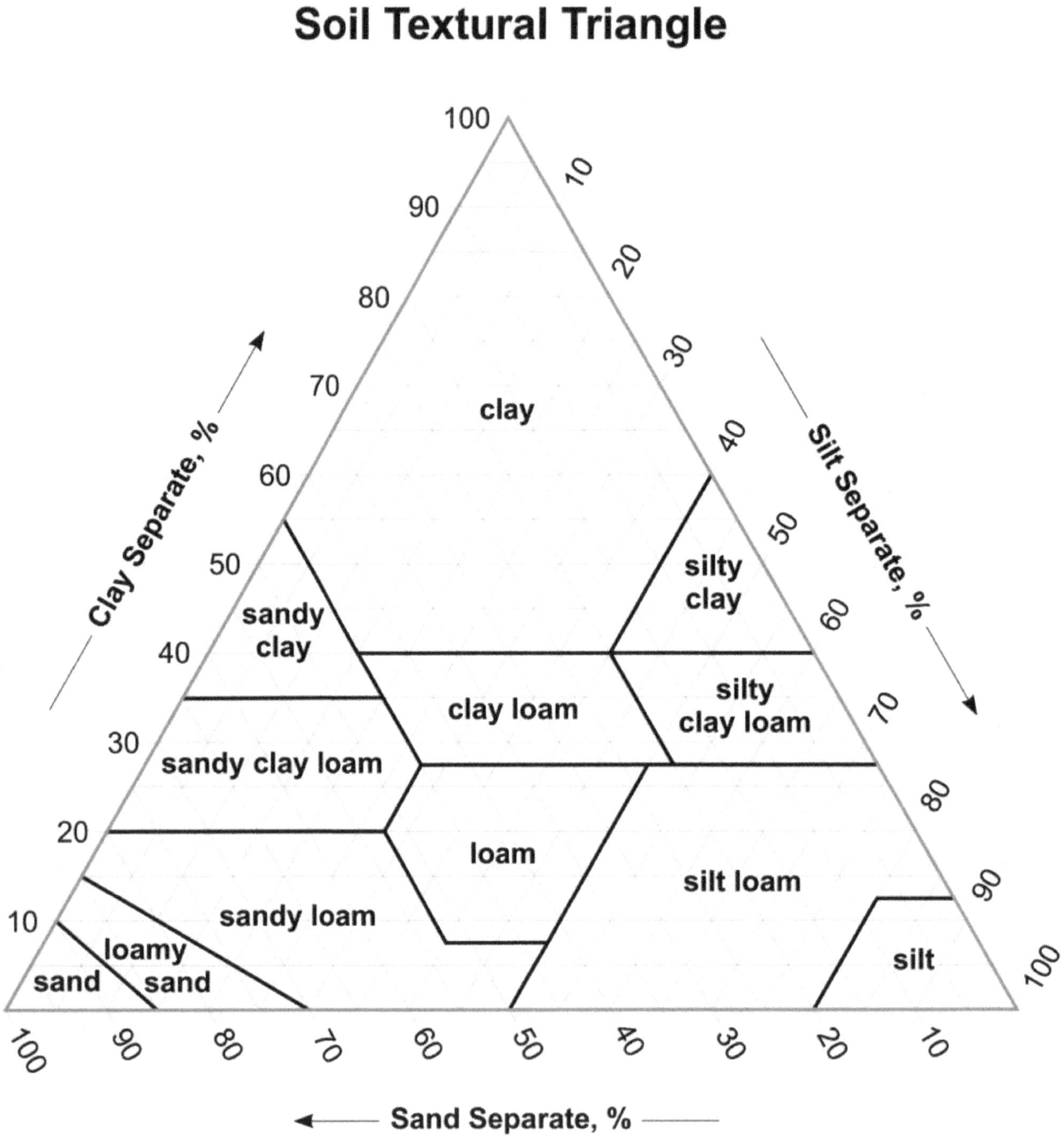

Figure 7.2. Soil texture triangle. Diagram courtesy of the USDA-NRCS.

Activity 2: Estimating Soil Texture by Feel

A soil scientist often needs to estimate soil texture while in the field or when laboratory data on the amounts of sand, silt, and clay are not available. With practice, you can learn to estimate texture by simply feeling or manipulating a moist sample.

To learn this technique, consider a simplified and generalized version of the textural triangle (Fig. 5.3). This modified triangle consists of three tiers based on approximate clay content. Clays are very cohesive, plastic, and can be easily molded. Clay loams are intermediate in clay content, cohesiveness, and ease of molding. Loams are soils low enough in clay content to possess little cohesiveness and are more difficult to mold. Sands do not form stable forms when molded.

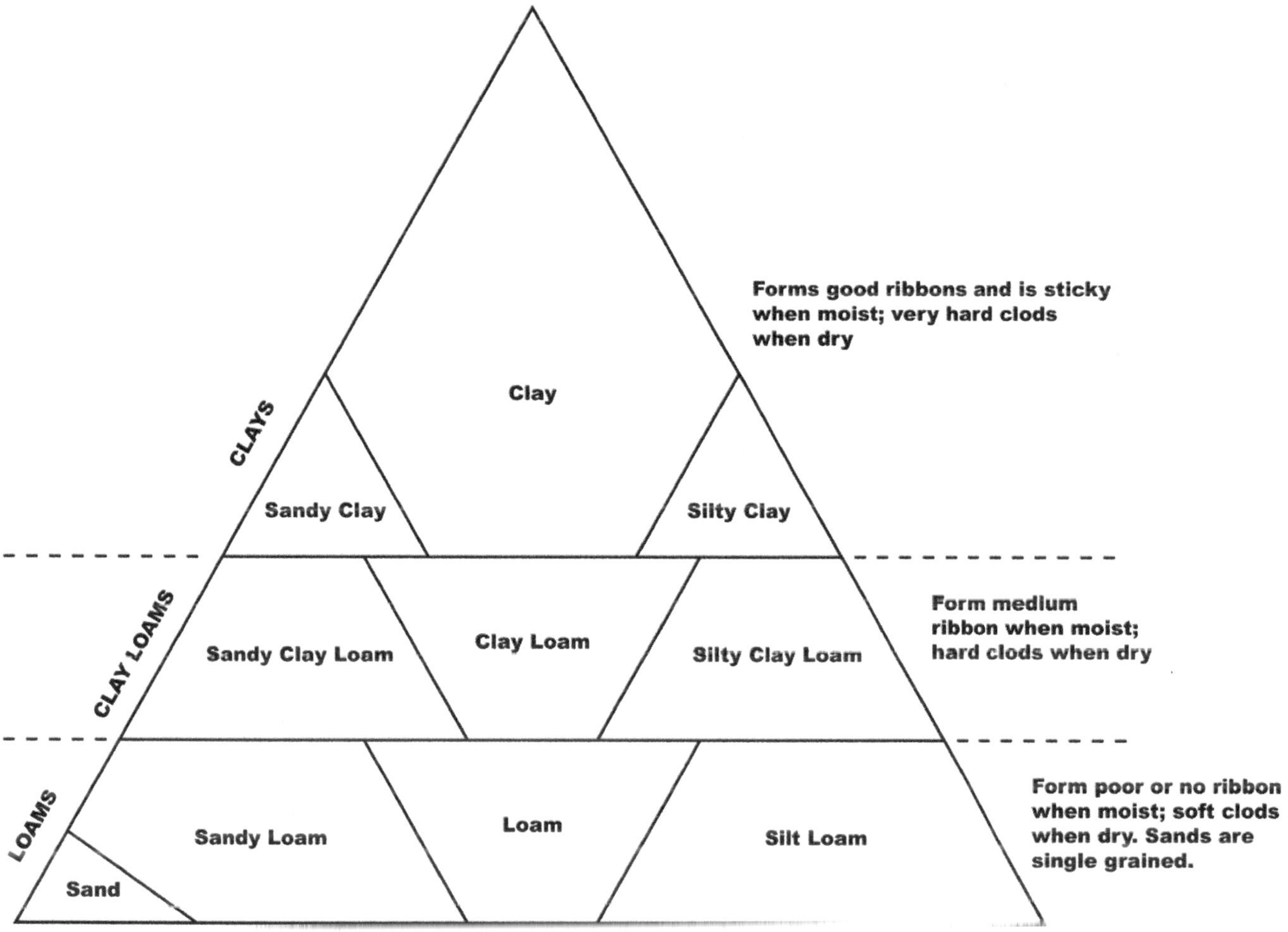

Figure 7.3. Modified texture triangle for determining soil texture by feel. The three tiers based on clay content are subdivided based on sand or silt content. Where neither sand nor silt is dominant, a prefix is not used in the name. Diagram courtesy of King et al. (2003).

Using the procedure outlined in Figure 7.4, determine the texture on the samples provided in the laboratory. Working the sample at the proper moisture content is very important. The sample must be moistened throughout. Achieving the correct moisture condition may take several minutes. After moistening and mixing soil to the proper consistency, perform the ribbon test, the grittiness test, and the smoothness test as described in the diagram. Try to estimate the texture of the samples, then check the answers provided. After you have calibrated your fingers on the practice samples, determine the texture for the unidentified samples provided, and enter your estimate in Table 7.3.

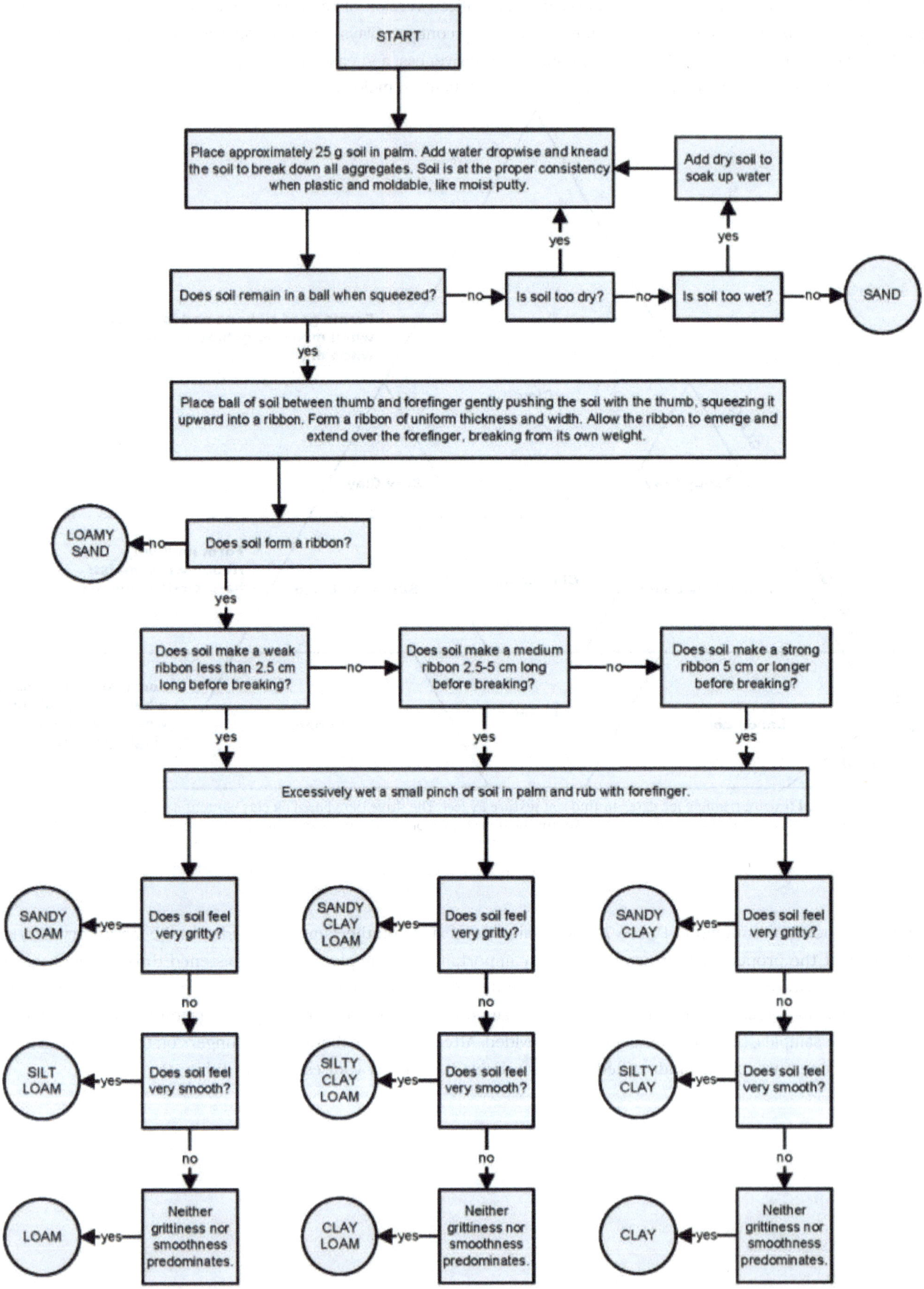
START
Place approximately 25 g soil in palm. Add water dropwise and knead the soil to break down all aggregates. Soil is at the proper consistency when plastic and moldable, like moist putty.
Add dry soil to soak up water
Does soil remain in a ball when squeezed?
Is soil too dry?
Is soil too wet?
SAND
yes
no
Place ball of soil between thumb and forefinger gently pushing the soil with the thumb, squeezing it upward into a ribbon. Form a ribbon of uniform thickness and width. Allow the ribbon to emerge and extend over the forefinger, breaking from its own weight.
Does soil form a ribbon?
LOAMY SAND
Does soil make a weak ribbon less than 2.5 cm long before breaking?
Does soil make a medium ribbon 2.5-5 cm long before breaking?
Does soil make a strong ribbon 5 cm or longer before breaking?
Excessively wet a small pinch of soil in palm and rub with forefinger.
Does soil feel very gritty?
SANDY LOAM
SANDY CLAY LOAM
SANDY CLAY
Does soil feel very smooth?
SILT LOAM
SILTY CLAY LOAM
SILTY CLAY
Neither grittiness nor smoothness predominates.
LOAM
CLAY LOAM
CLAY

Figure 7.4. Texture-by-feel method for estimating soil texture. Flowchart courtesy of the USDA-NRCS.

For the unidentified samples, record the ribbon length, predominant wet feel (smooth, gritty, neither), and textural class names determined by the feel method shown in Table 7.3.

Table 7.4. Results for estimation of soil texture by feel.

	Does is make a ball?	Does it make a ribbon?	Ribbon Length	Predominate Wet Feel	Textural Class
1.					
2.					
3.					

Table from King et al. (2003).

Activity 3: Particle Size Analysis by Hydrometer

Particle size analysis is based on the principle that different size particles fall through a fluid at different rates.

A particle falling in a fluid is subjected to 3 forces: gravity, buoyancy, and friction (Figure 7.5). The gravity and buoyancy forces are constant, but the frictional force increases as the velocity increases (like the drag force on an airplane increases as it goes faster). Because of this increasing frictional force, the particle eventually reaches a constant velocity (terminal velocity). Constant velocity occurs when the sum of the forces acting on the particle is zero, or acceleration is zero.

Stoke's Law is derived by setting up an equation containing the three forces acting on the particle when acceleration is zero:

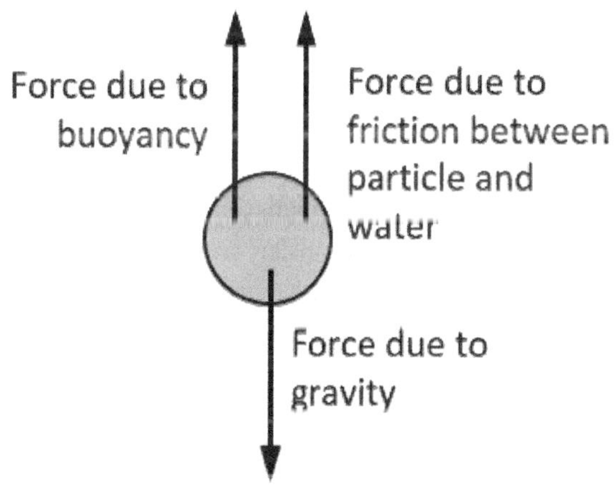

Figure 7.5. Forces acting on soil particles falling in water. Diagram courtesy of King et al. (2003).

$$\text{Force of gravity} = \text{force of buoyancy} + \text{force of friction}$$

These forces are determined from the following relationships:

Force of Gravity

$$\text{Force of gravity} = \text{mass of particle} \times \text{acceleration of gravity}$$

$$\text{Mass of particle} = \text{Volume of particle} \times \text{density of particle}$$

Therefore,

$$\text{Force of gravity} = \text{volume of particle} \times \text{density of particle} \times \text{acceleration of gravity}$$

Force of Buoyancy

$$\text{Force of buoyancy} = \text{mass of water displaced by particle} \times \text{acceleration of gravity}$$

$$\text{Mass of displaced water} = \text{volume of particle} \times \text{density of water}$$

Therefore,

$$\text{Force of buoyancy} = \text{volume of particle} \times \text{density of water} \times \text{acceleration of gravity}$$

Friction Force

Friction force is a function of the size of the particle, the velocity of the particle, and the viscosity of the water.

$$\text{Friction force} = 6\pi \times \text{viscosity} \times \text{radius of particles} \times \text{velocity}$$

After substituting the appropriate components of the forces into the equation, it can be solved for the terminal velocity:

$$\text{Velocity} = \frac{\text{particle diameter}^2 \times \text{acceleration due to gravity} \times (\text{particle density} - \text{liquid density})}{18 \times \text{viscosity of liquid}}$$

Note that the larger the diameter of the particle, the faster it settles (sand grains will settle faster than silt particles, which will settle faster than clay particles). Also, the density and viscosity of water vary with temperature, so the velocity of settling will be influenced by the temperature of the water (Figure 7.6).

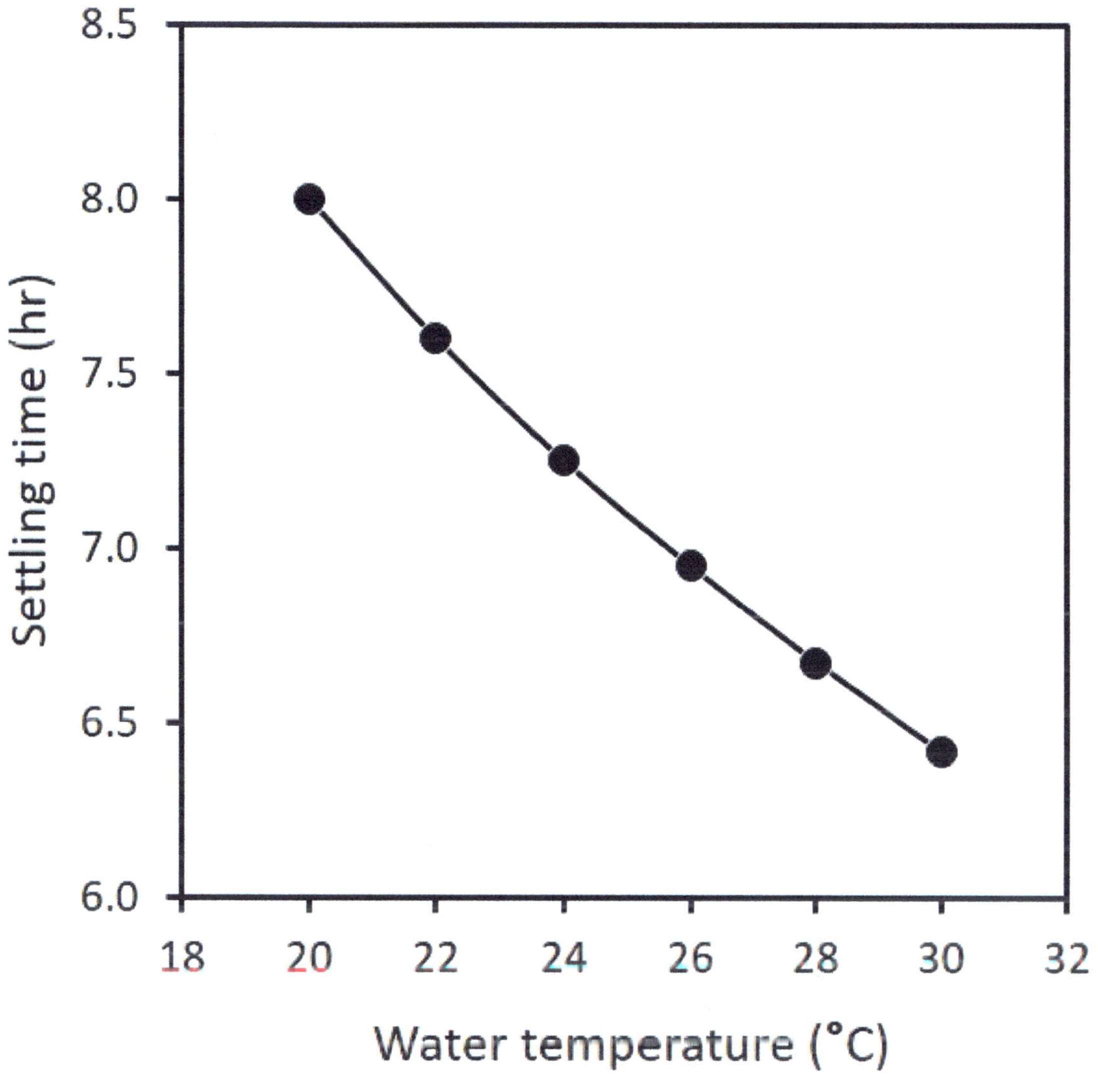

Figure 7.6. Time for a clay particle to settle 10 cm in water as influenced by water temperature. Graph based on King et al. (2003).

Stoke's Law can be simplified by specifying the temperature of the water:

$$\text{Velocity (cm s}^2) = \text{K} \times (\text{diameter of particle in cm})^2$$

Where K (cm^{-1} s^{-1}) is a constant incorporating water density and viscosity and acceleration due to gravity. Because velocity is distance/time, this equation can be solved for time required for a particle of a specified diameter to fall a given distance.

$$\text{Velocity} = \frac{\text{distance}}{\text{time}} = \text{K} \times (\text{diameter})^2$$

Therefore,

$$\text{Time} = \frac{\text{distance}}{\text{K} \times (\text{diameter})^2}$$

Thus, for a particle of a given diameter, the time required for the particle to fall a specified distance can be calculated.

Consider this example: How long will it take a 0.05-mm particle to fall 10 cm in water at 25°C?

At 25°C, K = 10,000 cm^{-1} s^{-1}. Substituting this value and the diameter (0.005 cm) into the above equation yields:

$$\text{Time} = \frac{10\ \text{cm}}{10000\ \text{cm}^{-1}\ \text{s}^{-1} \times (0.005\ \text{cm})^2} = \frac{10}{0.25\ \text{cm s}^{-1}} = 40\text{s}$$

Thus, after 40 seconds, the upper 10 cm of a soil-water suspension is completely free of all particles 0.05 mm or larger, so it is free of sand and contains only silt and clay.

Let's do the same calculation for a 0.002-mm particle, which is the upper limit of the clay range.

$$\text{Time} = \frac{10\ \text{cm}}{10000\ \text{cm}^{-1}\ \text{s}^{-1} \times (0.002\ \text{cm})^2} = \frac{10}{0.0004\ \text{cm s}^{-1}} = 6.94\text{hr}$$

Thus, after 6 hours and 56 minutes, the upper 10 cm of a soil-water suspension is free of all particles 0.002 mm or larger (sand and silt), so it contains only clay particles.

As we can see, Stokes Law can be used to determine when a volume of a soil-water suspension will be devoid of soil particles larger than a given size. Then we can measure the concentration of soil remaining in that volume. For example, after 40 seconds, we can measure the concentration of soil in suspension in the upper 10 cm of a suspension and thus determine how much clay + silt are present.

Under natural conditions, sand, silt, and clay particles are bound together in aggregates. These aggregates must be broken down so soil particles act independently of each other. For example, an aggregate of clay particles would behave as a silt particle, a phenomenon we want to avoid.

Dispersion is a 2-step, chemical/mechanical process. First, sodium hexametaphosphate (like the dishwashing detergent "Calgon®") is added to a soil-water suspension to increase electronegativity of soil clays; it causes a repulsive force between clay particles. Then the suspension is stirred vigorously (milkshake mixer or blender) to assure complete dispersion. The repulsive forces generated by the chemical treatment tend to stabilize the dispersed condition. Dispersion thus assures that aggregated clay particles do not behave like silt-sized or sand-sized particles.

Soil organic matter is an important binding agent, so it first must be removed by oxidation (using hydrogen peroxide, for example). In soils with very low in organic matter, this step is often omitted. It will not be used in this exercise.

After proper dispersion, sand, silt, and clay can be separated and quantified by allowing the particles to settle in water. (NOTE: a sieve is commonly used to quantify sand content of a soil sample, and then a hydrometer to quantify silt and clay content of the remaining particles. We will use a hydrometer for all three particle sizes in this lab activity.)

One method of determining the concentration of soil in suspension is using a hydrometer to measure the density of the suspension. The hydrometer is commonly used in field labs. In this exercise, we'll use a hydrometer calibrated to read directly in g/L of suspension.

Simplified Hydrometer Procedure

Chemical dispersion (performed in the previous lab)

1. Weigh out 30.0 g of dry soil (assume oven-dry) into a 250-ml Erlenmeyer flask.
2. Wash sides of flask with distilled water from a wash bottle.
3. Add 100 ml of distilled water using a graduated cylinder, and add 10 ml of sodium hexametaphosphate solution (500 g/L) from the dispenser on the sodium hexametaphosphate bottle.
4. Swirl to mix.
5. Wash sides of flask with distilled water from a wash bottle.
6. Cover the flask with Parafilm and label the flask with your lab section and table number as well as soil type. Store the flasks in the location specified by your instructor until next laboratory period.

After chemical dispersion

1. Quantitatively transfer the dispersed sample into a metal dispersion cup (a milkshake mixer cup), fill to half full with distilled water, and mix for 5 minutes as directed by your instructor. ("Quantitatively transfer" means to transfer all the sample.)
2. Quantitatively transfer the sample into a 1-liter sedimentation cylinder.
3. Fill cylinder to the 1000-ml mark with distilled water.
4. Suspend all soil particles in the sedimentation cylinder using one of the following methods:
    - Plunger method: Carefully insert stirring plunger and move up and down the full length of the cylinder for 30 seconds, ensuring that all particles are thoroughly mixed. Hold base of cylinder firmly with other hand.
    - Stopper method: Place a rubber stopper (one properly sized for the sedimentation cylinders) into the top of the cylinder, and holding both the bottom of the cylinder and the top of the stopper, mix the solution vigorously by inverting the cylinder and turning it right side up repeatedly for 30 seconds.
5. Record time when stirring is stopped and plunger removed, or when the stoppered cylinder is returned (quickly but gently) right-side-up to the laboratory bench.
6. Immediately insert the hydrometer slowly and carefully. Read the hydrometer (top of the meniscus) at exactly 40 seconds after stirring was stopped or the cylinder was returned to the bench. (Note: if a soil is high in organic matter that was not removed before beginning this experiment, bubbles will form at the surface. To disperse the bubbles, add three to five drops of alcohol to the surface of the solution immediately after inserting the hydrometer.
7. Repeat steps 4 through 6 until readings are within 0.5 units of each other. Record the reading on the data sheet.
8. Determine the temperature of the suspension and make corrections for temperature and for the blank as described in Table 7.4:
    - For each degree above 20°C, add 0.36 to the hydrometer reading.
    - For each degree below 20°C, subtract 0.36 from the hydrometer reading.
    - Because the soil suspension also contains sodium hexametaphosphate, each hydrometer reading must be corrected to account for the effect of sodium hexametaphosphate on density. A "blank" cylinder has been set up containing only water and sodium hexametaphosphate. Record the hydrometer reading from this cylinder and subtract that value from all hydrometer readings.
9. Your lab instructor has conducted measurements for the 7-hour reading and will provide the data to you to complete your soil texture determination.

10. Calculations:

Remember that these hydrometer readings are in g/L. Because the volume in the cylinder is one liter, the readings gives the amount of soil in the cylinder. For example, the 7-hour reading indicates the amount of clay (g) in the cylinder. Therefore, the percent clay is

$$\text{Percent clay} = \frac{\text{corrected 7 hr reading}}{\text{mass of dry sample}} \times 100$$

The 40-second reading is used to calculate percent silt + clay:

$$\text{Percent silt } + \text{ clay} = \frac{\text{corrected 40 s reading}}{\text{mass of dry sample}} \times 100$$

Then the silt and sand percentages can be determined.

$$\text{Percent silt } = (\text{percent silt} + \text{percent clay}) - \text{percent clay}$$

$$\text{Percent sand } = 100 - (\text{percent silt} + \text{percent clay})$$

Use Table 7.4 to help determine the textural class of your samples. Label all results in the proper units.

Table 7.5. Soil Texture by Hydrometer Method Data.

Row	Formula		Soil A	Soil B	Soil C
a		Dry weight of soil, g			
b		Average of 40-second hydrometer readings			
c		Hydrometer reading from blank			
d	b-c	Hydrometer reading corrected for blank			
e		Temperature of suspension, first readings			
f	d + [(e - 20) x 0.36]	40-second hydrometer reading corrected for temperature			
g		7-hour hydrometer reading			
h		Hydrometer reading from blank			
i	g-h	Hydrometer reading corrected for blank			
j		Temperature of suspension, second reading			
k	i + [(j - 20) x 0.36]	7-hr hydrometer reading corrected for temperature			
l	f ÷ a x 100	Percent silt + clay			
m	k ÷ a x 100	Percent clay			
n	l-m	Percent silt			
o	100 - m - n	Percent sand			
		Textural class (from triangle)			

Table adapted from King et al. (2003).

Soil Texture by Hydrometer Method Calculations

Activity 4: Soil Structure

Table 7.6. Summary of soil structure types described in mineral soils.

Structure Type	Aggregate Description	Usual Location
Granular	Relatively nonporous, small and spheroidal peds; not fitted to adjoining aggregates	A horizon
Platy	Aggregates are plate-like. Plates often overlap and impair permeability.	E horizon
Angular blocky	Block-like peds bounded by other aggregates whose sharp angular faces form the cast for the ped. The aggregates often break into smaller blocky peds.	B horizon
Subangular blocky	Block-like peds bounded by other aggregates whose rounded subangular faces form the cast for the ped.	B horizon
Prismatic	Column-like peds without rounded caps. Other prismatic aggregates form the cast for the ped. Some prismatic aggregates break into smaller blocky peds.	B horizon
Columnar	Column-like peds with rounded caps bounded laterally by other columnar aggregates that form the cast for the peds.	B horizon in alkali soils

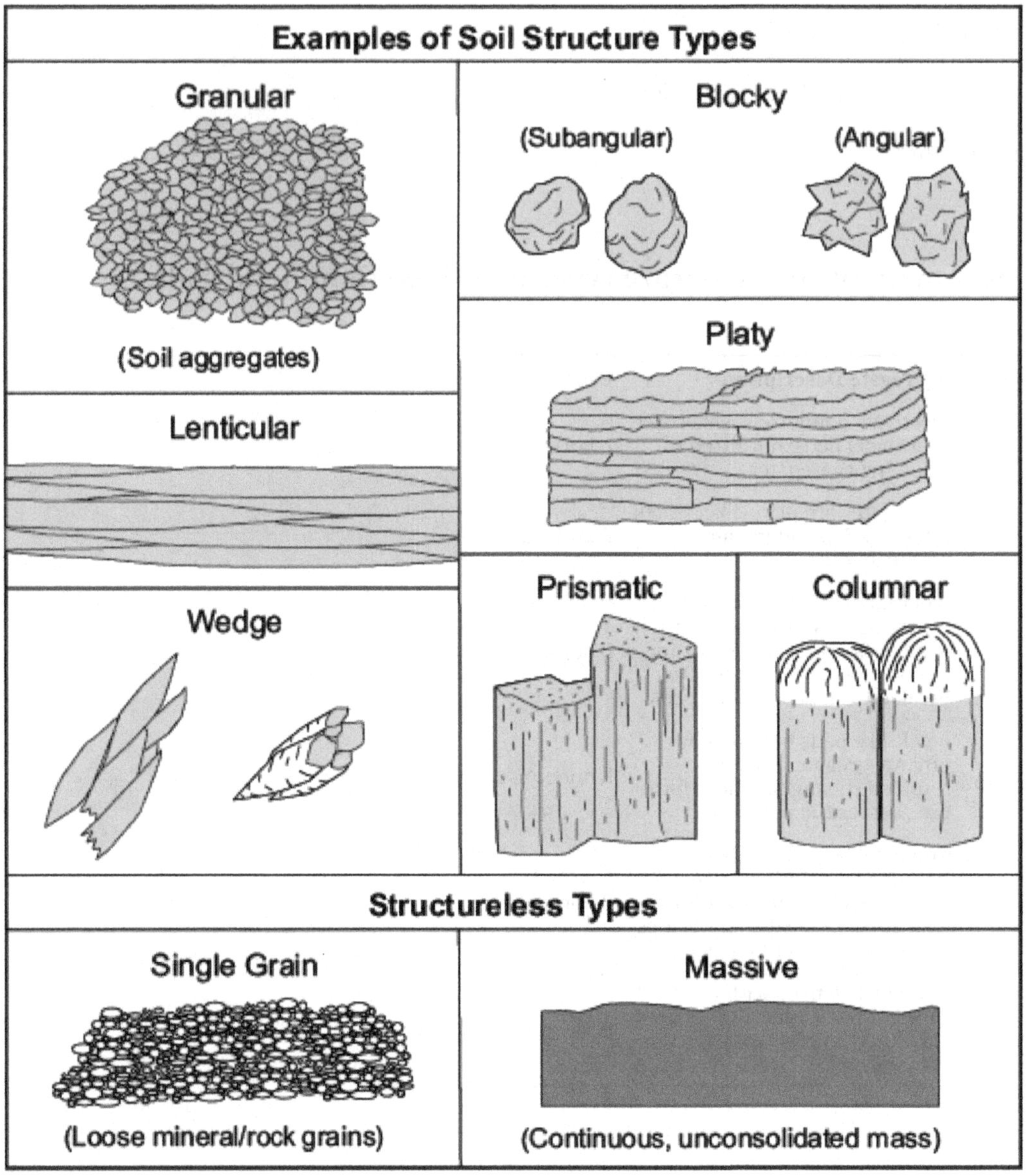

Figure 7.7. Examples of Soil Structure Types. Diagram courtesy of the USDA-NRCS.

Carefully examine the different structure types listed in Table 7.6 and shown in Figure 7.7, then answer the following questions.

- How does blocky structure affect permeability of clayey soils?
- How would a structureless massive condition affect permeability?
- How would manipulation (such as tillage) of a wet clayey soil affect structure?

Assignment: Online Quiz

A quiz will be available online. Please access it as directed by your instructor.

8. Soil Density

The flow of liquids and through soil is essential for the existence of plants and other soil organisms. Those flows are dependent on soil porosity and pore connectivity. Soil porosity is also dependent on, and inversely related to, soil density. If a soil is compacted, the soil solids are packed into a smaller volume, and the particles get packed closer together. This results in less total pore volume. Thus as soil density increases, soil porosity decreases. Soil density is relatively simple and cheap to measure. This laboratory entails a demonstration of the use of a soil density sampler, and a problem set that is focused on common soil density calculations.

Learning Objectives

- Measure soil bulk density and calculate pore volume relationships.

Materials

- Soil bulk density sampler
- Soil Density Problem Set

Recommended Reading & Viewing

- Soil Bulk Density Overview (CropWatch – Youth, 2013a)
- Soil Bulk Density Test (CropWatch – Youth, 2013b)
- Bulk Density (USDA NRCS, 2008a)
- Soil Bulk Density/Moisture/Aeration (USDA NRCS, 2019)

Prelab Assignment

Using the recommended reading and viewing resources and the introduction to this lab, answer the questions listed below. These definitions/questions will provide a concise summary of the major concepts addressed in the lab. They are also useful as study notes for exams.

1. Define and explain the differences among particle density, bulk density, and specific gravity.
2. Describe the relationship between 1) bulk density and porosity and 2) particle density and porosity. Write a mathematical expression for these relationships.
3. What is the typical range of bulk density values for mineral soils? What is the range of bulk density values for an organic soil?
4. What are the sizes of macropores and micropores?
5. Briefly describe the processes of soil structure (aggregate) formation.
6. Describe several examples of soil management practices that increase or decrease soil bulk density.

Introduction

Soil physical properties control the mechanical behavior of soils and will strongly influence land use and management. Several important physical properties have been discussed in other labs: texture, structure, color, and consistency. In this laboratory, additional physical properties will be measured, and their implications for land management will be discussed.

Bulk Density, Particle Density, and Pore Space

Soil represents a unique arrangement of solids and voids. The voids, or pore space, are important for air and water movement and storage. The total pore space consists of the voids between sand, silt, and clay particles and voids between soil aggregates. Therefore, texture and structure govern the amount of soil pore space. Organic matter affects the solids portion of the soil but also influences porosity indirectly through its effect on structure.

Density refers to a mass per unit volume. *Bulk density* of a soil refers to the mass of a volume of dry soil. The volume includes both solids and pores. *Particle density* refers to the mass of solids per volume of the solids alone. These two density measurements provide an important insight into the physical nature of a given soil. Soil density plays a major role both in plant growth and in engineering uses of soil. The formulas for calculating bulk density and particle density follow:

$$\text{Bulk density, } \rho_b = \frac{\text{mass of oven dry soil}}{\text{total soil volume}}$$

$$\text{Particle density, } \rho_p = \frac{\text{mass of oven dry soil}}{\text{volume of soil solids}}$$

Units of density are typically expressed in g cm^{-3} or Mg m^{-3}.

Note that total volume of the soil sample equals the volume of the solids and the volume of the pores.

Porosity is the ratio of the volume of the pores in a soil sample to the total volume of the sample:

$$\text{Porosity, } \varphi = \frac{\text{volume of pores}}{\text{total soil volume}}$$

However, measuring the volume of pores in a soil sample is difficult. In practice, porosity is normally calculated using the formula:

$$\text{Porosity, } \varphi = 1 - \frac{\rho_b}{\rho_p}$$

Porosity is usually expressed as a decimal, but it can also be expressed as a percentage by multiplying the decimal form by 100%.

Samples for determining bulk density must be collected very carefully to insure the sample represents the in situ condition desired and no additional compaction or loosening has occurred. One method for determining bulk density is the "core" method. A (relatively) undisturbed, cylindrical soil core is collected using a device like the one shown in Figure 8.1.

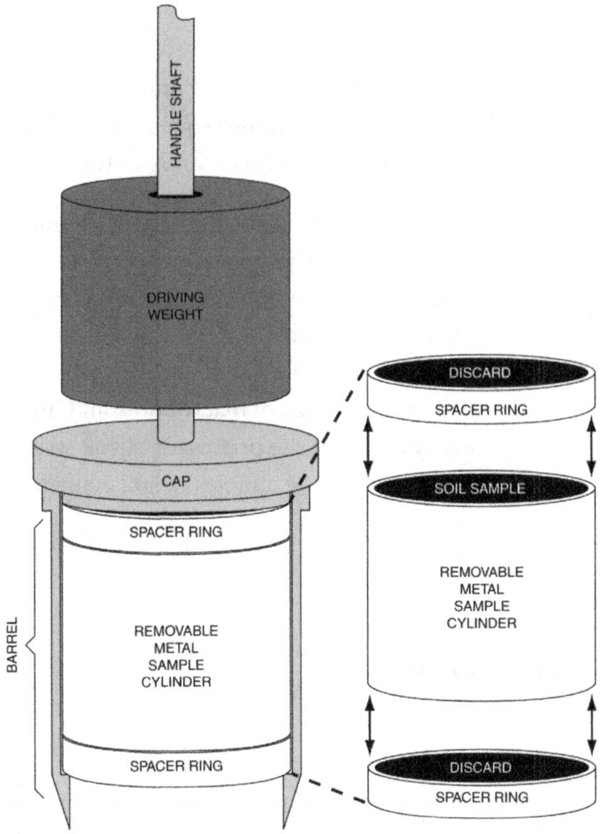

Figure 8.1. Uhland sampler for collecting soil samples for bulk density determinations. Diagram from King et al. (2003).

The driving weight is raised and dropped repeatedly to drive the sampler into the soil. When the desired depth is reached, the device is removed from the soil, and the removable metal cylinder containing the soil sample is removed. With the dimensions of the cylinder and the weight of oven-dry soil inside the cylinder, we can calculate the bulk density.

Sampling Procedure

1. Insert a 1.5 cm metal ring, a 6 cm metal core, and then a second 1.5 cm metal ring into the barrel of the core sampler, then reattach the barrel to the handle. The sample will be held in the longer cylinder; the two 1.5 cm rings are spacers, which help ensure an undisturbed soil sample.
2. Place the sampler over the desired sampling location, and then drive it into the soil with the slide hammer at the top of the handle. Stop when the cap of the barrel is flush with the soil surface.
3. Remove the sampler from the soil by pushing against the handle until the vertical shaft of the handle is parallel to the soil surface.
4. Remove the barrel from the sampler and gently push the core out of the top of the barrel, taking care to keep the core intact.
5. Carefully cut between the two shorter rings and the main core. Place the core into a labeled, pre-weighed canister, and put on the lid.
6. Back in the lab, weigh each canister plus moist soil.

7. Determine the weight of dry soil in the sample.
8. Measure the length and diameter of the metal cylinders.
9. Use this information to calculate bulk density, porosity, and water-filled pore volume.

Determining the weight of dry soil in the sample

The simplest method is to dry the sample in a conventional oven:

1. Remove the lids of all of the canisters, and place each in a 105C oven. Dry overnight.
2. Record the final weight of each canister (including the lid) plus oven-dry soil.
3. Calculate the moisture content of the samples:

$$\text{Mass of water} = (\text{mass of beaker} + \text{moist soil}) - (\text{mass of beaker} + \text{dry soil})$$

$$\text{Mass of dry soil} = (\text{mass of beaker} + \text{dry soil}) - \text{mass of beaker}$$

$$\text{Percent moisture} = \frac{\text{mass of water in soil}}{\text{mass of oven dry soil}} \times 100\%$$

1. Calculate the dry weight of the soil in each cylinder and record the data:

$$\text{Dry weight} = \frac{\text{wet weight}}{1 + (\frac{\text{percent moisture}}{100\%})}$$

Activity 1 & Assignment: Problem Set

The problem set will be provided to you at the beginning of the laboratory session. You will be working with your classmates to complete it during lab. Your lab instructor will help you as needed.

9. Soil and Water Relationships

Recent major droughts around the U.S. demonstrate the importance of the efficient use of water. Accurately measuring the amount of water available to plants is essential in determining irrigation rates and reducing the amount of wasted water. Soil water content is typically measured in the field using soil moisture sensors, which will be discussed in more detail in the Soil Water Measurement and Movement lab. Results from any soil moisture sensor must be interpreted using a "soil water relationship". Soil water relationships are graphs depicting a measured value versus soil water content, and can be considered a calibration specific to a particular soil. In this lab, soil water relationships will be developed for a clayey soil and a sandy soil.

Learning Objectives

- Understand the effect of texture on pore size and water holding capacity.
- Know relationships among moisture potential, movement, and availability.
- Determine soil water content by weight and volume.
- Measure saturated hydraulic conductivity and understand factors influencing rates of water movement.
- Explain the operation of moisture monitoring devices.

Materials

- Two soils, a sandy soil and a clayey soil, both air-dried, ground, and passed through a 2 mm sieve
- Shallow trays capable of holding water to a depth of 5 cm
- Plastic cups
- A tool (such as a thumbtack) to punch holes in the bottom of the plastic cups
- Filter paper or paper towels
- Scissors
- Permanent markers and label tape
- Laboratory balances accurate to the nearest 0.01 g
- Weigh boats
- Soil scoops
- Graduated cylinders
- Oven
- Soil drying tins

Recommended Reading

- Module 5 – Plant Water Relations (Allison and Jones, 2005)
- Soil, Water, and Plant Relationships (Rogers et al., 2014)
- Available Water Capacity (USDA NRCS, 2008b)

Prelab Assignment

Using the recommended reading resources and the introduction to this lab, consider the questions listed below. These definitions/questions will provide a concise summary of the major concepts to be addressed in the lab. They are useful study notes for exams.

1. Define soil water potential in your own words.
2. Define and compare gravitational water, plant available water, and unavailable water. Relate each to the different pore sizes (macro-, meso-, and micro-).
3. Identify the matric potential values (soil water potential) for saturation, field capacity, wilting point, air dry, and oven dry soil.
4. Compare and contrast soil water content on a weight (mass) basis and a volume basis. Show how to convert between these weight and volume.
5. Common units used to quantify moisture potential are cm (of water), bars, and kPa. How many cm of water in one bar? How many kPa in one bar? How many cm of water in one kPa?

Introduction

An understanding of soil-water relationships is essential to proper soil management for agronomic or other land use purposes. The soil-water-air continuum involves principles of physics, chemistry, and biology. Water, which can limit plant growth, is probably most important in evaluating land use, influencing soil characteristics for most land uses.

The retention and movement of water in soils, its uptake and translocation in plants, and its loss to the atmosphere are all *energy-related* phenomena. The energy relationships between free water and soil moisture are generally expressed in terms of soil-water potentials. These potentials are described in units of pressure, generally kilopascals (kPa).

Water content and water potential are related as shown in Figure 9.1 for five soils of differing textures. In general as clay content increases, the soil water content increases for a given soil moisture potential. The reason for this is that clayey soils have a higher porosity, and can hold on to more water at a given soil water potential.

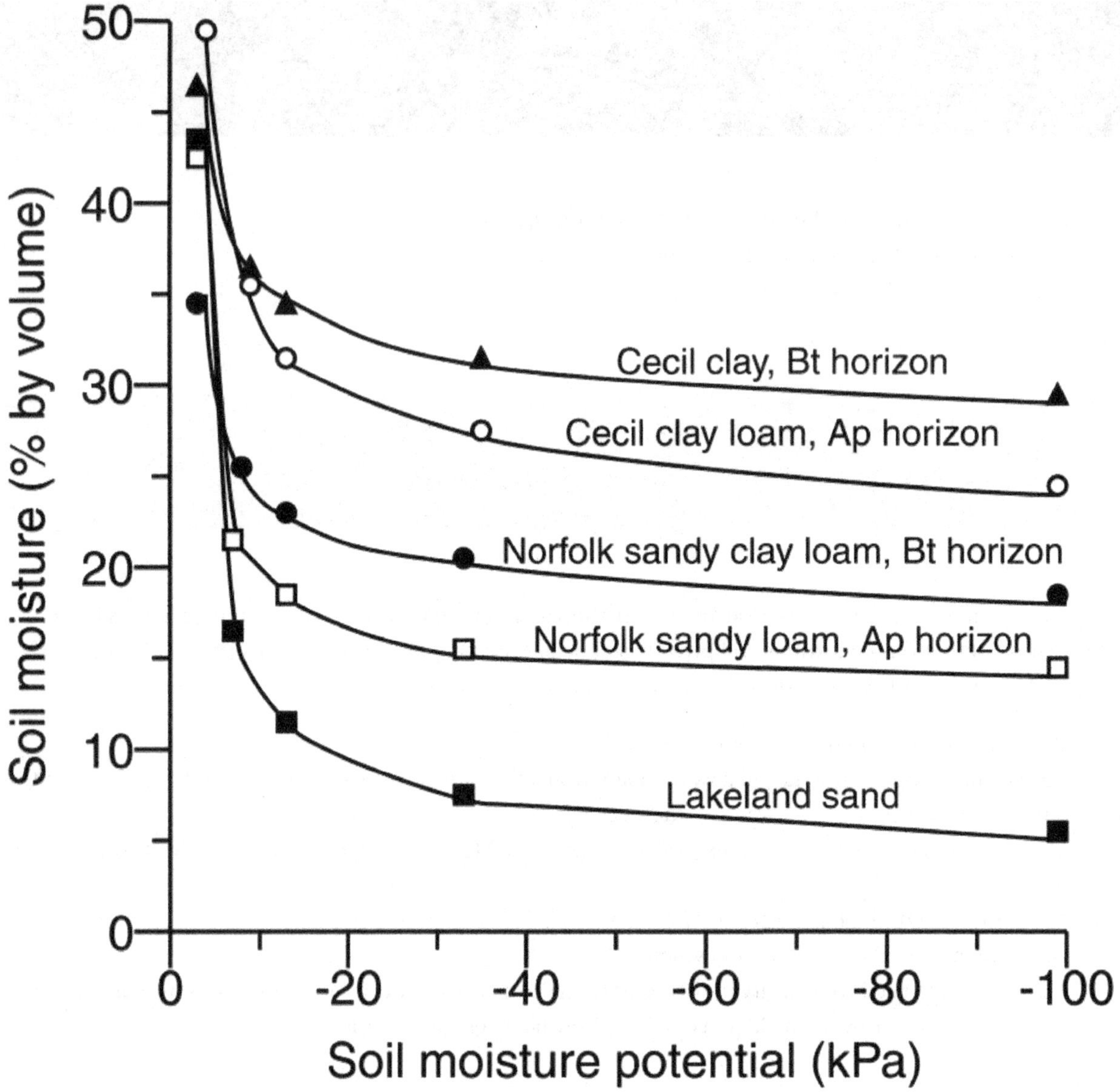

Figure 9.1. Relationship between moisture content and moisture potential for three soils down to -100 kPa. Graph from King et al. (2003).

The relationship between soil moisture content and soil moisture potential is a continuum. However, to understand moisture availability to plants, we break the continuum into parts and give each part a name or classification, as shown in Figure 9.2 and explained in Table 9.1. The important points on the continuum include saturation, field capacity, wilting point, air dry, and oven dry. Therefore, for a plant, field capacity and wilting point are most important because they determine the amount of plant-available water.

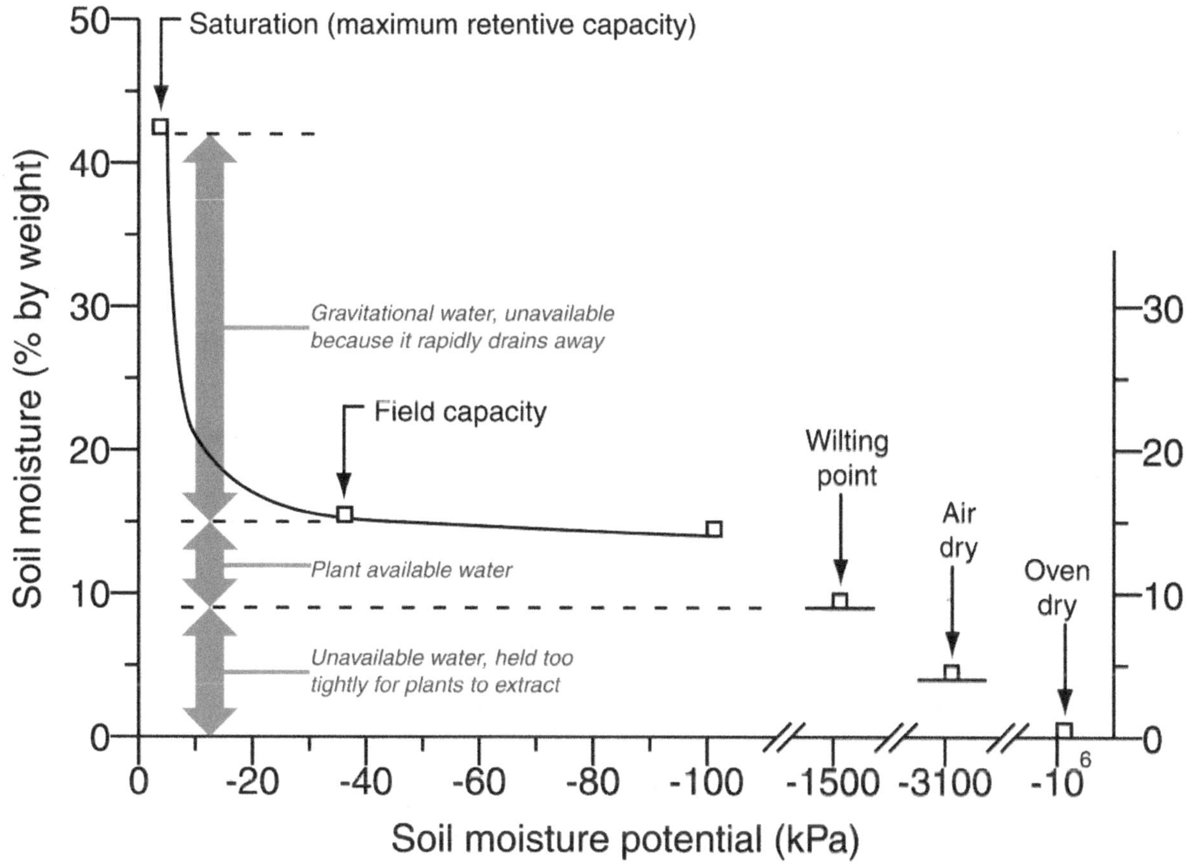

Figure 9.2. Soil moisture classes and important points on the soil moisture relationship curve. Graph based on King et al. (2003).

Table 9.1. Classes, calculations, and explanations of soil water.

Classes of Water	Calculation and Explanation
Gravitational water	water content at saturation – water held at field capacity
	water that drains from macropores too rapidly for plants to use
Plant available water	water held at field capacity – water held at wilting point
	water in mesopores
Unavailable water	water held at wilting point
	mostly hygroscopic water held too tightly in micropores to be plant available

Methods of Expressing Soil Water Content

The moisture content of a soil can be expressed using either weight or volume. Moisture content measured using weight is determined by drying a moist sample to a constant weight. The difference in weights before and after drying represents the soil water. The percentage is determined as follows:

$$\text{Percent soil water by weight} = \frac{\text{mass of water}}{\text{mass of oven dry soil}} \times 100\%$$

Remember to divide by the oven dry weight, not the wet weight.

This approach is called gravimetric analysis and yields the gravimetric water content.

Volumetric moisture content is determined as follows:

$$\text{Percent water by volume} = \frac{\text{volume of water}}{\text{volume of soil}} \times 100\%$$

The gravimetric and volumetric water contents do not equal each other. However, we can use bulk density to convert from weight to volume and vice versa.

Recall that bulk density is:

$$\text{Bulk density} = \frac{\text{mass of oven dry soil}}{\text{volume of soil}}$$

Also recall that the density of water is 1.0 g/cm^3, i.e., a 1.0 g of water has a volume of 1.0 cm^3.

Multiplying soil moisture measure by weight x bulk density yields:

$$\frac{\text{Mass of water}}{\text{Mass of oven dry soil}} \times \frac{\text{Mass of oven dry soil}}{\text{Volume of soil}} = \frac{\text{Mass of water}}{\text{Volume of soil}} = \frac{\text{Volume of water}}{\text{Volume of soil}}$$

For example, a soil with a bulk density of 1.5 g/cm^3 and a water content of 30% measured using weight will have a volumetric water content of 45% (30 x 1.5 = 45).

The volumetric water content is a useful measurement. First, it more realistically describes the moisture environment of growing plant root systems as they explore a given volume of soil. Secondly, volumetric water values can be converted to represent water content at a particular equivalent soil depth or thickness. Representing water at a specific depth changes the expression from a volume (cm^3) to a linear (cm) expression of soil water:

$$\text{Depth of soil} \times \text{ water content by volume} = \text{depth of water}$$

For example, a 25 cm layer of soil with a volumetric water content of 20% (or 0.20 when expressed as a decimal fraction) would contain 5 cm of water in the 25 cm layer of soil:

$$25 \text{ cm soil} \times 0.20 = 5 \text{ cm of water}$$

Activity 1: Moisture Content Calculations

Two days after a heavy soaking rain, a soil sample was collected. During two days between rainfall and sampling, the gravitational water drained out of the soils, so the soil moisture content was at field capacity. After 10 days of hot dry weather, plants on the soil began to wilt, and a second sample was collected. These two soil samples were weighed, oven-dried, and re-weighed. The soil had a bulk density of 1.3 g/cm^3. The following data were collected:

Table 9.2. Soil moisture content data

Sample	Moisture state	Moist weight	Oven-dry weight
1	Field capacity (2 days after rain)	160 g	128 g
2	Wilting point (10 days after rain)	170 g	156 g

Calculate the following quantities. Calculations from Sample 1 data are shown as an example.

What is the percent water by weight (gravimetric water content) in each sample?

$$\text{Sample 1 percent water by weight} = \frac{(160\text{ g} - 128\text{ g})}{128\text{ g}} \times 100\% = 25\%$$

$$\text{Sample 2 percent water by weight} =$$

What is the percent water by volume (volumetric water content) in each sample?

$$\text{Sample 1 percent water by volume} = \frac{1.3\text{ g}}{\text{cm}^3} \times 25\% = 33\%$$

$$\text{Sample 2 percent water by volume} =$$

What percent water by weight is plant available (difference between field capacity and wilting point)?

How many centimeters of plant available water could be held in a 50-cm layer of this soil?

When the soil water content is at the wilting point, how deeply would a 3 cm rainfall wet this soil? (Assume no runoff.) HINT: Room for storage of water without runoff would include water volume up to field capacity. Therefore, the difference in volume between wilting point and field capacity would indicate how deeply the soil can absorb rainfall.

Activity 2: Measuring Soil Moisture at Saturation, Field Capacity, and Wilting Point

Steps 1-10 below were completed in a previous lab. The objective of this activity is to examine the influence of texture and density on water holding capacity and pore size distribution, the moisture status of different soil materials will be monitored through a drying cycle. Each lab group will be responsible for two soils and one treatment (AD, WP, FC, and S). Your laboratory instructor will assign the treatment. You will complete the experiment on both soils.

1. Label an empty cup as follows:
    - Lab Section number
    - Soil Type: Sandy or Clayey

- Treatment:
    - AD(air dry)
    - WP(wilting point)
    - FC(field capacity)
    - S(saturated)

2. Punch eight small holes in the bottom of the cup.
3. Cut a piece of paper towel and place in the bottom of the cup to prevent soil from falling through the holes.
4. Weigh the empty cup and paper towel, and record the value on the data sheet.
5. Place a cup on the balance, tare out the weight of the cup, and add exactly 50g of soil. Gently tap the cup on the lab bench to settle the soil.
6. We need to know the volume of soil in the cups so we can calculate bulk density. Fill another cup (without holes!) with water to the same level as the soil in the cups. Pour the water into a graduated cylinder and record the value on the data sheet (assume 1 ml = 1 cm^3).
7. Get the data for the other soil from the lab groups with the other soil.
8. Calculate and record the bulk density.
9. Carefully and slowly place the WP, FC, and S cups in the water in the dishpans to allow the soil to become saturated from the bottom to the top. Your lab instructor will remove WP cups tomorrow and allow them to dry for 6 days. The FC cups will be removed in 6 days and allowed to drain for 1 day. The S cups will remain in the water all week.
10. Place the AD cup in the dishpan with no water.

Now complete the exercise by weighing all cups of soil. Subtract the weight of the empty cup, and record on the instructor's master datasheet. The instructor will provide a composite of the data on the blackboard for your section following experiment summary. A lab worksheet will be handed out in class, so you can complete calculations based on the data your lab group compiled. Complete this worksheet, and turn it in at the following lab period.

Experiment Datasheet						
Weight of empty cup + towel [g]		____________g				
Weight of soil [g]		____________g				
			Sandy Soil		Clayey Soil	
Volume of water in cup (i.e., volume of soil)			____________cm^3		____________cm^3	
Bulk Density			____________g/ cm^3		____________g/ cm^3	
			Sample Weight			
			Sandy Soil	Sandy Soil	Clayey Soil	Clayey Soil
Treatment	Code	Moisture class	Cup + Soil	Soil	Cup + Soil	Soil
Dried in oven	OD	Oven dry				
Never Wet	AD	Air dry				
Drained 6 days	WP	Wilting point				
Drained 1 day	FC	Field capacity				
Never Drained	S	Saturated				

Assignment: Lab Summary

For this lab, you will be preparing a lab summary. The assignment will be provided to you at the beginning of the lab.

10. Soil Water Measurement and Movement

In the Soil and Water Relationships lab we developed soil water relationships for a sandy soil and a clayey soil. These soil water relationships are essential for interpreting results from any soil water sensor. Once calibrated, those sensors can be used to quantify plant available water. They can also be used to determine the direction of water flow due to gravity, capillarity, or even osmotic potential. While it's possible to control an irrigation system from a smart phone, a thorough knowledge of the principles that govern water measurement and movement is essential to effectively employing such technology. In this lab, we will watch a classic video on soil water movement, participate in some hands-on activities related to soil water content, and observe some state of the art soil moisture sensors.

Learning Objectives

- Become familiar with the driving forces of soil water movement and how soil properties can affect soil water movement.
- Review the key soil water contents and types of water held in soil at different tensions.
- Examine some modern instruments used to measure soil water content and describe how they function.
- Review common calculations for soil water measurement and movement.

Materials

- Soil Water Measurement and Movement Problem Set
- Computer with internet access and a projector
- Tensiometer with a vacuum gauge or pressure transducer
- WATERMARK 200SS electrical resistance blocks with user manual (Irrometer Company, Inc., Riverside, CA, U.S.)
- EC-5 Small Moisture Sensor with user manual (EC-5 Small Moisture Sensor, Decagon Devices, Pullman, WA, U.S.)
- Hydra Probe Soil Sensor with user manual (Stevens Water Monitoring Systems Inc., Portland, OR, U.S.)

Recommended Reading

- Module 4 - Water and Solute Transport in Soils (McCauley and Jones, 2005b)

Prelab Assignment

Using the recommended reading, the Soil and Water Relationships lab, and the introduction to this lab, consider the questions below. These definitions/questions will provide a concise summary of the major concepts to be addressed in the lab. They are useful study notes for exams.

1. Describe why fine-textured soils are suitable for dryland farming while coarse textured soils are not.
2. Rank soils with the textures clay loam, loam, and sandy loam from the soil with the highest water tension to the soil with the lowest water tension.
3. Define a soil moisture retention curve (also known as a soil water relationship), and describe how it is used.
4. Describe why fertilizer placement is important in irrigated systems.

Introduction

Soil water movement is an important process in soil; it controls the amount of water available to plants, how much water can be stored in the soil, and whether the root zone has excess water in. Soil water movement is classified into saturated flow, unsaturated flow, and vapor flow.

Saturated flow occurs at tensions of 0 to -0.3 bar, between saturation and field capacity. The rate of flow depends on hydraulic conductivity, which is controlled by pore size. In general, the coarser the material, the faster the flow rate. The driving force behind saturated flow is hydrostatic head, so the higher the column of water, the more hydrostatic head. This is like diving to the bottom of a pool. The deeper you go, the larger the column of water above you. With enough pressure, your ears can pop.

Unsaturated flow, also known as capillary flow, occurs between 0.3 to 31 bars, between field capacity and air dry. In general, soil water moves through capillary action from areas with the most potential energy, to areas with the least potential energy. Essentially, it moves from wet areas to dry areas. In other words, the drive force is a tension gradient. Hydraulic conductivity controls how fast the water may flow. Under unsaturated conditions, hydraulic conductivity is controlled by tortuosity, or how direct or indirect the path of water flow is. A straight path is the most direct and thus the fastest route, while the least direct path with be the slowest route.

Vapor flow is like capillary flow but is driven by gradients in vapor pressure instead of water tension.

Activity 1: Water Movement in Soils Video

Watch the Water Movement in Soils video (Hsieh et al., 1961) as a class, and then answer the following questions:

- What two forces are responsible for the movement of water upward against the downward force of gravity?

- When water is first added to the irrigation furrow, the movement of water outward (to the side) is ________ the movement of water downward.

    A. Less than

    B. Equal to

    C. Greater than

- When do gravitational forces predominate?

- For the demonstration using a fine soil with a layer of coarse material (sand), describe what happens initially when the wetting front reaches the sand.

- For that same demonstration, what must occur for the water to pass into the coarse layer?

- For the same demonstration, what happens when the saturated layer reaches the sand?

- For the demonstration with a coarse soil containing a layer of fine material (such as clay), describe what happens initially when the wetting front reaches the clay.

- For that same demonstration, why does a water table form above the clay layer?

- What happens when free water is applied directly to a layer of coarse material? What is the driving force behind the rapid movement?

- What happens when a layer of coarse material is not in direct contact with free water?

- In the comparison of water movement in a sandy loam, loam, and a clay loam, which texture has the deepest penetration of irrigation water? Why?

Why is dryland farming practical on fine-textured soils, but not on coarse-textured soils?

Describe how soluble fertilizers move within the furrow in relation to both the irrigation channel and the crop root system?

Where should tile drains be placed so they carry away excess water?

Activity 2: Soil Water Movement Sponge Demonstration

Using the sponge and tray at your table, squeeze the sponge while it's submerged under water.

Equating the sponge and the water it holds to soil and soil water, what water content does this sponge have?

A. Saturation

B. Field capacity

C. Wilting point

D. Air dry

The water that freely drained away from the once-saturated sponge is considered what type of water?

A. Gravitational water

B. Plant available water

C. Unavailable water

D. Capillary water

Now, pull the sponge out of the water, hold it above the pan, and let it drip. Notice that water will freely drain from the sponge for a while but eventually stops.

Equating the sponge and the water it holds to soil and soil water, what water content does the sponge have now?

A. Saturation

B. Field capacity

C. Wilting point

D. Air dry

Apply slight pressure to the sponge to remove some water, and let the water drip into the pan. Notice that as you apply more force, more water is removed from the sponge. This resembles how plants remove water from the soil. Initially very little energy is exerted for them to extract water. However, as the soil dries, more and more energy is required to remove water.

Equating the sponge and the water it holds to soil and soil water, what water content does the sponge have once you can no longer remove any water by squeezing the sponge?

A. Saturation

B. Field capacity

C. Wilting point

D. Air dry

Notice that the sponge is still moist. This is because even beyond the point where plants can remove water from the soil, some water still remains. This water is called what?

A. Gravitational water

B. Plant available water

C. Unavailable water

D. Capillary water

Activity 3: Devices for Measuring Soil Moisture

In the textbook, review the devices for measuring soil moisture and the principle upon which each device operates. Observe the devices on display in the lab.

Tensiometer

What does the meter actually read?

Does the tensiometer measure soil moisture content or soil moisture potential?

Electrical Resistance Blocks and Watermark-brand Sensors

What does the meter actually read?

How is moisture content determined from the meter reading?

Volumetric Water Content Sensor (Decagon EC-5)

What does the meter actually read?

How is moisture content determined from the meter reading?

Stevens Hydra Probe Soil Sensor

What does the meter actually read?

How is moisture content determined from the meter reading?

Activity 4 & Assignment: Problem Set

The problem set will be provided to you at the beginning of the laboratory session.

PART IV
SOIL BIOLOGY

11. Compost Facility Field Trip

Composting is essentially decomposition under controlled conditions. It uses soil organisms to break down organic matter, resulting in CO_2 being released to the atmosphere, and ultimately resulting in organic matter that's highly concentrated in nutrients. It is performed at various scales for many reasons, such as municipalities composting food and yard waste to reduce landfill expenses, farmers composting deceased animals to reduce rendering costs, gardeners turning their autumn leaves into valuable mulch, and a coffee drinker feeding coffee grounds and filters to earthworms. In this lab we will examine the Soils Teaching Lab worm farm, and visit a model industrial scale composting facility at the Agronomy North Farm where dining center food waste and campus greenhouse waste is turned into compost to be later used for agricultural research experiments.

Learning Objectives

- Discuss vermicomposting and review how worm farms are managed.
- Visit the K-State composting operation at the Agronomy North Farm.
- Learn about managing a thermophilic composting operation.
- Discuss how organic waste from the KSU campus is turned into compost.
- Conduct a test for compost maturity

Materials

- Active worm farm
- Compost thermometer
- Basic Compost Maturity Test, Woods End Laboratories, Inc.
- Optional: Digital Color Reader – Multi-Mode Unit, Woods End Laboratories, Inc.
- Compost Maturity Test Directions, Woods End Laboratories, Inc.

Recommended Reading & Viewing

- Agronomy Compost Facility (Moorberg, 2020) (Note: *this is a video tour of the compost facility, and is an abbreviated version of the in-person tour.*)

- Types of Composting and Understanding the Process (US EPA, 2015)
- Solvita Compost Test: How to Perform It (Woods End Laboratories, 2009)

Prelab Assignment

Using the recommended reading and the introduction to this lab, consider the questions listed below. These definitions/questions will provide a concise summary of the major concepts addressed in the lab. They will also serve as the basis for the post-lab quiz and are useful study notes for exams.

1. Define vermicomposting.
2. Define thermophilic composting and describe how it results in the building up heat.
3. Name and describe the three stages of thermophilic composting.
4. Organic matter is decomposed during composting, resulting in a decrease in the overall C content and relatively low C:N ratios. Describe how the C:N ratio changes over time, and identify the gas that is released as C is emitted to the atmosphere.
5. What is the range of C:N ratios that would be expected in mature compost?
6. Why does compost need to be turned during thermophilic composting?

Introduction

Composting is the practice of creating soil organic material from decomposing organic matter under aerobic conditions outside the soil. The product, compost, is highly desirable, commonly used in potting soil mixes, as mulch, as slow-release fertilizer, and as a soil amendment. Compost can be produced under ambient temperatures through vermicomposting (composting using litter-dwelling worms) or via slow decomposition.

Vermicomposting is growing in popularity. If managed properly, worm farms are odorless, and the worms will rapidly turn fruit and vegetable waste and some paper products into nutrient-rich worm castings. Worm farms can be as small and simple as a plastic bin tucked under a desk, or large, industrial-scale operations. The Soils Teaching Lab worm farm is built with multiple levels. Note the increasing level of decay as you go from the top tray to the bottom. Common food wastes and some paper wastes make great substrates for worms. We will discuss what foods are ideal for worms, and what things should be avoided later.

A more common method of composting is thermophilic decomposition. Decomposing organic matter in thermophilic composting undergoes a dramatic buildup in heat due to both the mass of the decomposing material and the insulating properties of the pile itself. Thermophilic composting involves three distinct stages: the mesophilic, thermophilic, and curing stages. During the mesophilic stage, easy-to-decompose organic materials are metabolized, with a resulting gradual increase in temperature to more than 40°C. The next stage is the thermophilic stage, in which the temperature increases to between 50 and 75°C. Materials like cellulose that decompose more slowly and under more difficult conditions decompose during this stage. Regular mixing is required during this stage so that aerobic decomposition can be maintained. Without adequate oxygen, the humification process will slow, and the overall time required for decomposition will increase. The final stage is the curing stage. During this stage, the temperature drops to near-ambient conditions. As decomposition and humification progresses, carbon is removed from the organic matter and released to the atmosphere as CO_2. This removal of carbon decreases the C:N ratios and increases the percentages of N, P, and other nutrients in the compost.

At the K-State compost facility at the Agronomy North Farm, food waste from the campus dining halls and plant and soil waste from the campus greenhouses are broken down into compost that is then used for erosion, greenhouse, and field experiments. Figure 11.1 shows the C:N ratios, nitrogen content, and phosphorus content of both fresh and cured compost. The figure shows cattle feedlot manure for comparison. As carbon is removed through respiration and released to the atmosphere as CO_2, the C:N ratio decreases, and nitrogen and phosphorus content increase to levels comparable to the manure.

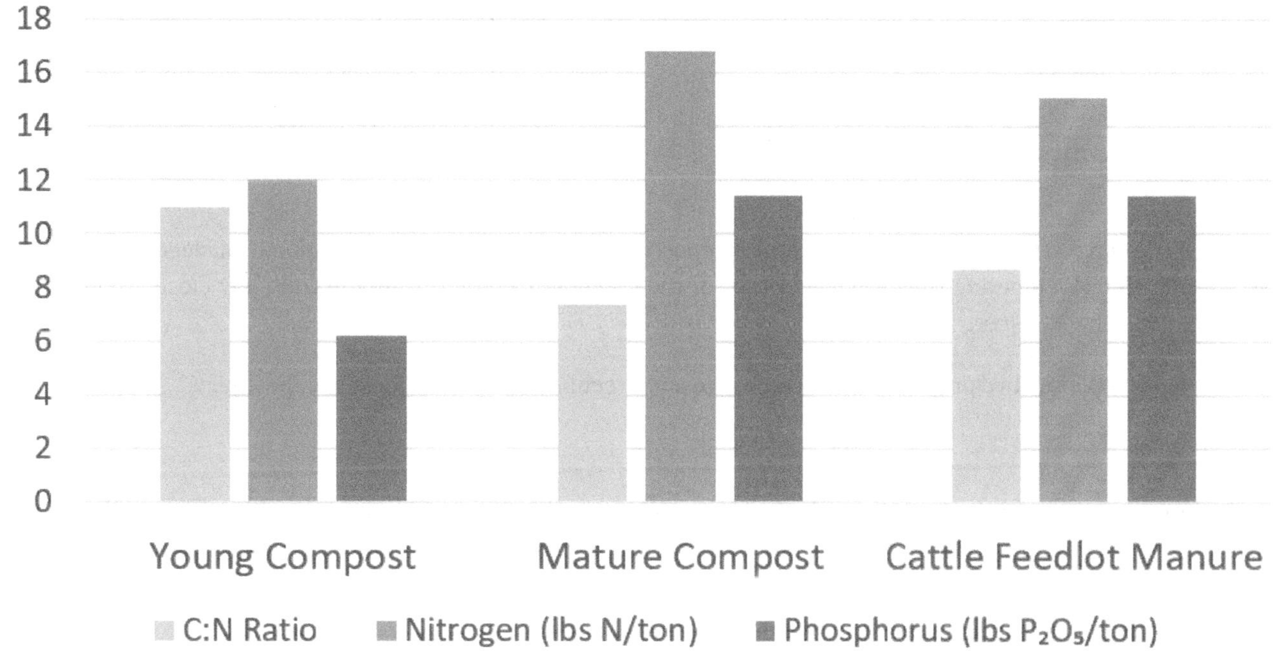

Figure 11.1. Nutrient analysis of young and mature compost and cattle feedlot manure. Data courtesy of D. Presley. Figure courtesy of C. J. Moorberg.

Composting using thermophilic decomposition organic waste is becoming very popular because of its many benefits. Some examples include:

- Organic materials are safely stored with minimal odor release.
- Compost is easier to handle than raw materials because of a 30 to 60% reduction in volume
- Nitrate depression during the decomposition of organic materials with high initial C:N ratios is limited to the compost pile and thus avoids any effect on plant growth.
- High temperatures during the thermophilic stage kills most weed seeds and pathogens.
- Toxic organic compounds like pesticides are destroyed by the time the compost is mature.
- Many soil-borne plant diseases are suppressed through microbial antagonism
- Because compost is made from recently photosynthesized carbon, it is carbon-neutral and thus a much more environmentally friendly choice than peat for potting mixes.

Activity 1: Vermicomposting Demonstration.

We will start with a group discussion of vermicomposting, and a demonstration of the Soils Teaching Lab worm farm. Participate in the discussion, and answer the following questions.

Why are redworms preferred over nightcrawlers?

What can foods and substrates can be added to worm farms? Why is gritty material needed?

What should not be added to worm farms?

Activity 2: Compost Facility Field Trip.

You will be traveling to the Agronomy North Farm compost operation. The compost operation is managed by Dr. DeAnn Pressley and Brent Wehmeyer. They will lead a tour of the compost facility. During the tour, pay close attention, and answer the following questions.

What are the relative proportions of food waste and greenhouse waste in these compost piles?

How often is the fresh compost pile turned? How often are the mature compost piles turned?

List the temperatures for each of the different compost piles, from the freshest pile (closest to the dumpsters) to the most mature pile (furthest from the dumpsters).

What practices are used to control unwanted animals on the compost pile, such as crows?

How long does it take for the compost to reach maturity?

How many pounds of waste are typically composted each year?

Activity 3: Testing for Compost Maturity.

The last thing you'll do at compost facility is collect samples from the compost piles to test for maturity. Collect one sample from a compost pile selected by your instructor using a clean trowel, then place it in a Solvita® test jar. Gently tap the jar on the ground or a bench top to allow it to compact to the compost's natural density. Add enough compost to reach the fill line. Then open the foil packs for a CO_2 test probe and a NH_4^+ test probe and place those probes in the jar so that the color can be viewed from outside the jar. Close and tighten the lid to allow the CO_2 and NH_4^+ gasses to react with the test probes. Mark the start time of the test on the jar lid, then allow the test to proceed for four hours at room temperature (20-25°C or 65-75°F). Your instructor will complete the test for you, and will send you a picture of the final test probes next to the color chart for reading the test probes. Your instructor may also provide you with more precise results analyzed using a Digital Color Reader to measure the color of the test probes.

- What is the number on the color chart for the color most closely corresponding to the color on the CO_2 test probe?

- What is the number on the color chart for the color most closely corresponding to the color on the NH_4^+ test probe?

- You will be provided a copy of the directions for the Solvita Compost Maturity Test. Based on the information provided in those directions, was the compost pile your lab group sampled mature?

- Compare the maturity of the various compost piles using the results from this maturity test?

Assignment: Online Quiz

A quiz for this lab will be available online. Please access it as directed by your instructor.

Solvita® is a trademark of Woods End Laboratories, Inc.

12. Soil Carbon and Respiration

Soil respiration is the net release of CO_2 from all living organisms in the soil, including bacteria, fungi, protists, earthworms, plant roots, etc. The rate of soil respiration has many different applications. It can be used assess soil health, with higher respiration rates indicating a more active, and thus healthy microbial community. It can also be used as part of a C balance to determine if soil under particular management practices is a net source or sink of C, which has direct application to carbon credits. Respiration rates can also be influenced by the quality of the substrate being consumed by the soil microbes. In this lab you will conduct tests to determine the relative soil health of two soils that have been managed differently for over a decade. The tests include a CO_2 burst or a basal respiration test to determine the relative respiration rates, and a slake test to determine the stability of the soil aggregates. You will also work with your classmates to complete the Carbon Cycling problem set.

Learning Objectives

- Define soil health.
- Describe how soil health indicators are used to track changes in soil health over time.
- Describe how soil health indicators are used to compare two different management practices.
- Explain how soil organic matter influences aggregate stability.
- Perform carbon cycle calculations for a Kansas farm.

Materials

- Two or more soil samples that have been oven-dried at 35-40°C (for the CO_2 burst test)
- Two or more moist soil samples that have been recently collected (for the basal respiration test)
- Soil roller
- 30 cc (cm^3) soil scoop and striker rod
- Plastic beaker
- Mesh screen
- Pipette set to 9 mL
- Water
- Solvita® CO_2 burst test paddle
- Test jar with lid
- Solvita® low-CO_2 test paddle
- Incubator (optional)
- Solvita® Digital Color Reader and/or reference color charts
- Two or more soil peds that are approximately the size of a golf ball

- Coarse wire mesh or machine cloth
- Two, 1 L sedimentation columns
- Soil Carbon Cycling Problem Set

Recommended Reading & Viewing

- Soil Respiration Overview (CropWatch – Youth, 2013e)
- Soil Health Demonstration: Slake Test (Montana NRCS, 2020)
- Soil Respiration (USDA-NRCS, 2009)
- Soil Respiration (USDA-NRCS, 2014b)
- Solvita® Soil CO_2 Respiration Updated Protocol Improvement (Woods End Laboratory, 2019)

Prelab Assignment

Using the recommended reading and videos and the introduction to this lab, consider the questions listed below. These definitions/questions will provide a concise summary of the major concepts addressed in the lab. They also useful study notes for exams.

1. Define soil health
2. Define soil respiration
3. Define carbon sequestration.
4. Name three ways in which C is lost from the soil.
5. Which of those three losses of C from the soil is the most significant?
6. What practices might reduce or even reverse this loss in soil C, and thus maintain or improve soil health?
7. Describe how soil organic matter impacts the stability of soil aggregates.

Introduction

Carbon Calculations

Carbon dioxide is a product of respiration, one of several greenhouse gases emitted from soils. Greenhouse gases adsorb radiation emitted from the earth into space, reflecting the radiation to the earth, thus heating the earth in an effect commonly called the "greenhouse effect". This process does help keep the planet's climate at a suitable temperature, but if the concentration of greenhouse gases in the atmosphere is too high, this process is magnified, trapping too much heat and resulting in warmer global temperature averages and changes to local climates. This can result in significant changes that affect people around the world. Soils are important in moderating greenhouse gases and climate change.

As microbes in the soil consume organic matter, they release carbon dioxide through respiration, which eventually makes its way to the soil surface and into the atmosphere. Undisturbed ecosystems, generally show a balance of C added to the soil and C lost from the soil to the atmosphere. In some cases, C accumulates in the soil when additions to C exceed C losses, resulting in soil features like a dark A horizon with lots of soil organic matter. This net increase in soil organic matter is called C sequestration. In disturbed soils, like tilled and cultivated soils, the physical disturbance of tillage temporarily increases aeration, resulting in higher rates of respiration and more C lost to the atmosphere. Over time, this causes a net decrease in soil C and net increases in atmospheric C.

However, improved soil management practices like conservation tillage, no-till, or using cover crops can minimize these C losses or even increase soil C. Soil management for C was a significant topic in the 2015 United Nations Climate Change Conference in Paris. There, scientists recognized that agriculture can both be a source of C emissions and as a tool to fight climate change through C sequestration by helping limit the average global temperature increase to 1.5°C. Carbon credits can be used to subsidize farmers who change their management practices to increase the amount of C sequestered (stored) in the soil in perpetuity. Carbon credits function as agreements between a C emitter and someone who can manage land so that C is removed from the atmosphere through C sequestration. In activity one, you will perform the calculations that a hypothetical Kansas corn producer would do when considering an opportunity to sequester C through changes to tillage practices and then sell C credits.

Measuring Soil Respiration

Soil respiration is a measure of how much CO_2 the soil is emitting and reflects the total CO_2 emitted from all living organisms in the soil, including bacteria, fungi, earthworms, protists, roots, and others. Soil respiration measurements are used for calculating carbon balances, and can be used as a biological indicator of soil health. In Activities 2 and 3, your class will work together to quantify respiration rates of soil samples that have previously been collected and dried (Activity 2) or samples that have recently been collected and are still moist (Activity 3). Your class will also investigate the aggregate stability of soils using a slake test (Activity 4).

Activity 1: Carbon Cycling Problem Set

A soil problem set will be provided to you. Use your experience and knowledge from previous laboratory activities to complete these problems.

Activity 2: Soil Respiration using Solvita® CO_2 Tests

Note: This protocol is an abbreviated version of the protocol that is provided with sampling kits from Solvita®. The full Solvita® will be provided to students to complete this activity.

Basal versus CO_2 Burst Tests

The CO_2 burst test is used in regions where prolonged periods of dry conditions are common. The CO_2 burst test simulates the respiration rate after it is rewetted following a prolonged dry period, and uses soils that have previously been collected and dried. Samples used for the CO_2 burst method must be rewetted. The basal test is used in more humid, temperate regions were soils rarely experience prolonged periods of dry conditions. The basal test uses soil samples that are minimally disturbed and have been recently collected. No additional moisture needs to be added to soils for the basal test. Your instructor will inform you which of these two tests you will be performing.

Procedure

For this experiment you will be using two similar soils preferable from the same area and of the same soil series. However, they should come from locations that have undergone contrasting soil management practices. One example is a soil that has been tilled annually compared to a soil that has been under no-till management and/or cover cropped. Another example is a tilled field compared to a soil from an adjacent, untilled fence row.

1. Collect the soil using a clean trowel, shovel, or probe. Collect 12 or more sub-samples from a representative area to the depth, then composite the sub-samples into a single sample. If you are conducting the basal respiration test, skip to Step 4.
2. Dry the soil samples in an oven at temperatures up to 45° oven until the soil reaches a steady state mass (several days to a week)
3. The soil should pass through a 2 mm sieve. If the aggregates are too large, roll the samples using a soil roller until the aggregates are broken down enough to pass through the sieve. Do not use a soil grinder.
4. Measure out a standard volume of soil using the 30 cc soil test scoop. Scoop enough soil so that soil is mounded on the top of the scoop. Then scrape off the excess sample using the striker bar, yielding exactly 30 cc of soil. Pour the sample into the plastic beaker and gently tap it to level out the sample. If you are conducting the basal respiration test, skip to Step 6.
5. Place the mesh screen on top of the soil in the beaker, then gently add 9 mL of water to the soil using a pipette.
6. Open the low-CO_2 pouch and place the test paddle into the soil with the stake pointing down and the color indicator vieweable on top of the sample.
7. Place the plastic beaker containing the soil sample and test paddle into the test jar, then tighly close the jar lid. Record the time the jar was closed on the jar lid. Keep the jar in a room at a stable 20°C for 24 hours, or place the jar in an incubator set to 20°C for 24 hours.
8. Remove the test paddle after 24 hours and record the resulting color by either comparing it to a color chart, or by measuring the color using a digital color reader (DCR). The DCR will provide both the color measurement and the interpolated CO_2-C ppm respired by the sample. If using the color chart, record the color number in Table 12.1, then use the "Range of Response Guidelines" from the Solvita® protocol to determine the CO_2-C ppm respired. Record results using a DCR in Table 12.2.

Table 12.1 Soil Respiration Results Using Color Chart

Soil	Color Measurement	CO2-C ppm Respired	Soil Biological Fertility Classes
A			
B			
C			

Table 12.2 Soil Respiration Results Using Digital Color Reader

Soil	Color Measurement	CO2-C ppm Respired	Soil Biological Fertility Classes
A			
B			
C			

Summarize the management practices for the soils as they where described by the instructor. Also briefly describe the soil (soil series, the horizon that was sampled, the depth of sampling, etc.).

Which test did you perform, the basal respiration test or the CO_2 burst test?

How did respiration from the soils compare?

Which soil is "healthier"? Explain the impact of soil management practices on the health of the soils.

Activity 3: Slake Test

Now you will conduct a slake test, which is a test that can be performed both in the field as well as in a laboratory. The slake test demonstrates how well a soil is able to withstand internal pressure that occurs during a sudden rewetting of the soil. Soils with more stable aggregates will be better able to withstand that pressure, and will exhibit less slaking than soils with less table aggregates. You will use two soils with contrasting soil management practices (see Activity 3). To perform the test, do the following:

1. Fill two, 1 L sedimentation columns with water.
2. Form a wire mesh into a "U shape" so that the two edges are supported by the rim of the column.
3. Place soil peds that are approximately the size of a golf ball onto the wire mesh. The peds should be completely submerged.
4. Observe the peds for at least 5 minutes.

Which soil exhibited the most slaking?

Describe how the history of soil management for the two soils compare, and explain how the soil management practices impacted this slake test.

Assignment: Problem Set

A problem set will be provided to you at the beginning of the laboratory session.

Subsequent Lab Setup

Some activities require more time than a single lab period. These activities must be set up ahead of time. One activity in the Soil Colloids explores the shrink-swell properties of soil. To prepare for this activity, do the following:

- In two 500 ml beakers, create slurries of kaolinite and bentonite clays (this should be sufficient for the whole class). The slurries should be approximately the same consistency as a milk shake: slightly runny, but not as thick as a Frosty®.
- You will be provided two petri dishes. Using a ruler, determine the interior volume of the dish. (Remember, volume of a cylinder = $h\pi r^2$, where h is the height of the cylinder, and r is the radius.)
- Using label tape, label the bottom of the two petri dishes with the following information:
    - Colloid type (Kaolinite or Bentonite)
    - Lab day/time
    - Table number
- Using a spatula, fill both petri dishes with the appropriate clay slurry.

- Set the petri dishes in the location designated by your instructor.

Solvita® is a trademark of Woods End Laboratories, Inc.

PART V

SOIL CHEMISTRY

13. Soil Colloids

Colloids consist of clay minerals and organic matter, and play critical roles in soil chemical, physical, and biological properties. Colloids are very small in size. The majority of charges in soil comes from colloids, making them important for cation exchange capacity, buffering capacity, and retention of nutrients like Ca^{2+}, K^{+}, Mg^{2+}, etc. In addition, the small size of colloids leads to very high surface areas, which facilitates chemical reactions and provides habitat for microbes. Some clays expand and contract with changes in soil wetness. The force from expanded clays can be considerable, and commonly cause structural failures in house foundations and basement walls, roads and bridges, and other man-made structures. Other clays do not expand, and make ideal substances for clay pottery, or tile roofs. The organic matter fraction of colloids are important for the development of soil structure, or as food sources for soil organisms. The soil properties attributed to colloids are numerous, and will be the focus of the following laboratory activities.

Learning Objectives

- Measure the effects of different cations on colloidal properties.
- Identify the soil components controlling ion exchange.
- Determine the cation exchange capacity of selected soil horizons.

Materials

- Petri dishes
- Beakers
- Spatulas
- Pure kaolinite
- Pure bentonite
- Caliper
- 6V battery
- Solid copper wires, 2 total, 25 cm long
- Alligator clips for the wires
- Paper towels
- Test tubes
- Flasks
- Burettes
- Bottle-top dispensers set to dispense 10 mL
- Phenolphthalein indicator solution
- Aluminum chloride solution, 1 M

- Potassium chloride solution, 1 M
- Sodium chloride solution, 1 M
- Calcium chloride solution, 1 M
- Sodium hydroxide solution, 0.01 M
- Simulated soil CEC extract solutions
- Norfolk E horizon, 0.002 M HCl
- 10 mL of extract represents 1 g of soil
- Norfolk Bt horizon, 0.005 M HCl
- 10 mL of extract represents 1 g of soil
- Cecil Ap horizon, 0.010 M HCl
- 10 mL of extract represents 1 g of soil
- Cecil Bt Horizon, 0.007 M HCl
- 10 mL of extract represents 1 g of soil
- Whitestore Bt horizon, 0.015 M HCl
- 10 mL of extract represents 0.5 g of soil

Recommended Reading

- Introduction to the Sorption of Chemical Constituents in Soils (Thompson and Goyne, 2012)
- Cation Exchange Capacity and Base Saturation (Sonon et al., 2017)

Prelab Assignment

Using the recommended readings and the introduction to this lab, consider the questions listed below. These definitions/questions will provide a concise summary of the major concepts to be addressed in the lab. They will also serve as the basis for the post-lab quiz and are useful study notes for exams.

1. Define the term "soil colloid". What are the main types of soil colloids?
2. Describe the sources of charges on soil colloids.
3. Define cation exchange capacity.
4. What units are used to express cation exchange capacity?
5. List the general quantity of cation exchange capacity contributed by kaolinite, montmorillonite (smectite), and humus.
6. Define percent base saturation. What cations are usually considered base cations? Technically speaking

they are not bases. Why are they called base cations?

Introduction

The extremely small, colloidal particles (smaller than 0.001 mm) of clay and humus control many important chemical and physical properties of the soil. This portion of the soil is often called the "active fraction". The small size of colloids results in a large surface area per unit weight, and their ionic structure results in a net electrical charge. The type, amount, and mineralogy of colloids will strongly influence most land management decisions. For example a soil that is 40% of clay that primarily consists of smectite (a 2:1 shrink-swell clay) could have limitations for constructing roads, or building foundations due to the shifting of the soil as the soil wets and dries. Such a soil could be highly productive for row crop agriculture though, due to the high amount of charge that facilitates the retention of nutrients like Ca^{2+}, K^+, Mg^{2+}, etc. On the contrary, a soil, such as an Oxisol that has 80% clay has colloids that are primarily aluminum and iron oxides, which do not shrink or swell, and have a low amount of charge. Thus, the soil would be well suited for building foundations. However, the high phosphorus fixation capacity, and the limited ability to retain base cations limit the productivity of the soil for row crop agriculture.

Ion exchange is one of the most significant features of the clay and humus fractions. The capacity of the particles to attract or adsorb cations is called the cation exchange capacity. This ability allows the soil to serve as a storehouse of plant nutrients like potassium, calcium, and magnesium. This reactive exchange capacity also permits the soil to serve as a filter or treatment medium for land application of waste materials.

Determining Cation Exchange Capacity

The cation exchange capacity (quantity of cations a soil can adsorb per unit weight, CEC) can be determined using a simple displacement process (Figure 13.1). In step 1, a soil sample is first saturated with a simple cation like NH_4^+ so all the negative charge sites are occupied by NH_4^+. In step 2, excess NH_4^+ (i.e., not on exchange sites) is removed by leaching with ethyl alcohol. In step 3, another cation such as Ba^{2+} is used to displace all the NH_4^+. The NH_4^+ is collected in the filtrate and measured. The quantity of NH_4^+ collected from the sample is the quantity of cations that the soil can hold, i.e. CEC.

Many cations could be used in step 1 of the displacement method. Soil laboratories often use the ammonium ion. However, in this activity, H_+ will be used as the saturating cation. Therefore, we will determine the amount of extracted H^+ in the filtrate. Then, using that amount, you will calculate the CEC for each soil sample.

Calculating CEC uses the concept of moles of charge. In chemistry, one mole of an element is the quantity of the element with a weight in grams numerically equal to its atomic weight. For example, the atomic weight of K is 39.1, so one mole of K weighs 39.1 g. One mole of any element contains 6.02×10^{23} atoms of the element. Similarly, one mole of charge is 6.02×10^{23} charges. If K exists as a cation in solution (K^+), then a solution containing a mole of K would contain 6.02×10^{23} positive charges.

The number of positive charges in the filtrate is thus equal to the number of negative charges on the exchange sites in the soil sample. Therefore, your task is to determine the number of charges in the filtrate. Each H^+ ion has one positive charge, so by determining the amount of H^+ in the filtrate, you can determine the number of positive charges in the filtrate and by extension, the number of negative charges on the exchange sites in the soil sample.

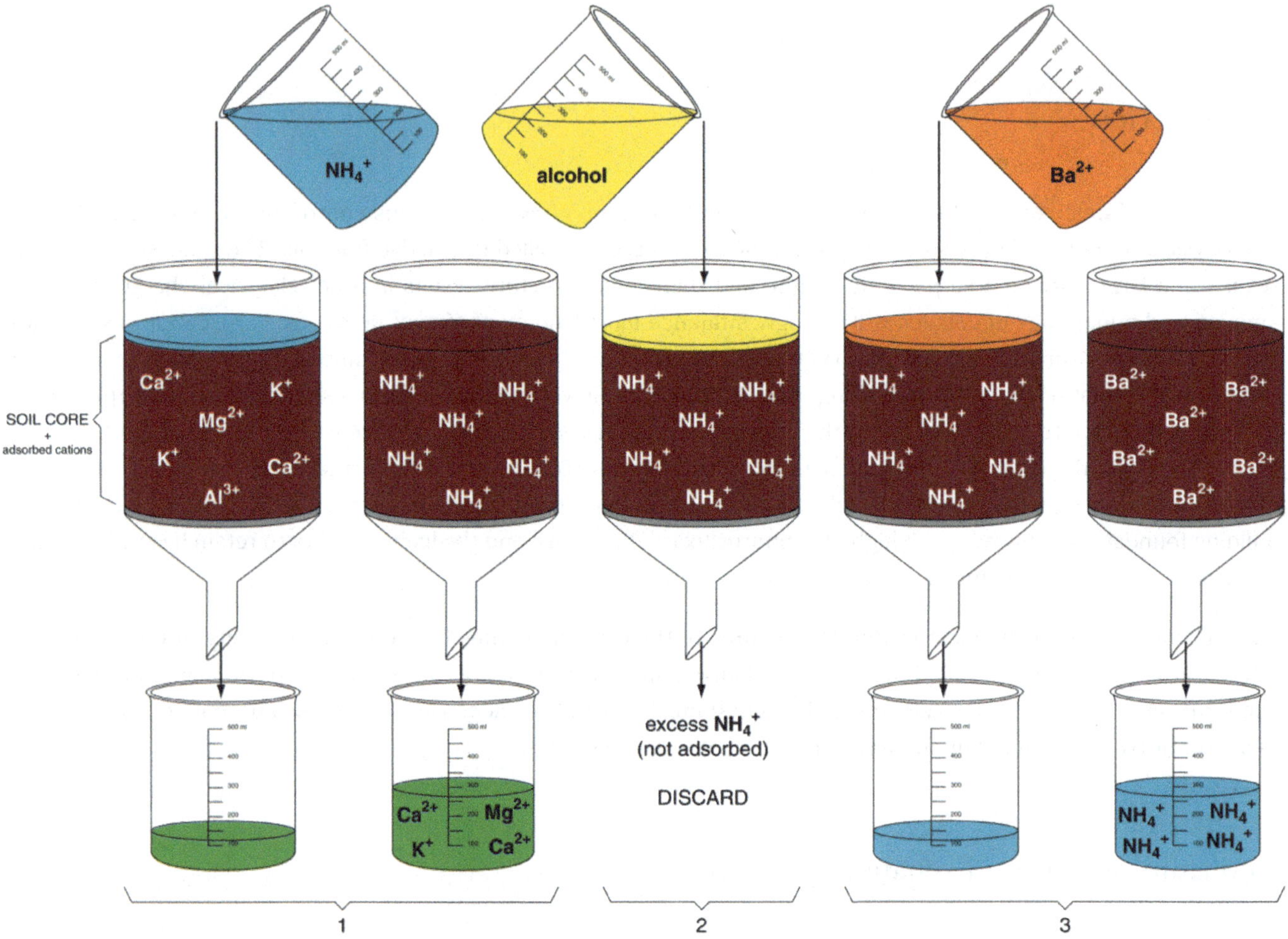

Figure 13.1. Series of chemical extractions to determine CEC. Diagram courtesy of King et al. (2003).

Activity 1: Demonstration of Shrink-Swell Characteristics

In a previous lab, you filled two petri dishes with slurries of bentonite and kaolinite and set them aside to dry. Retrieve those petri dishes and perform the following steps:

- Measure the height and radius of the discs of bentonite and kaolinite, and record the results in Table 13.1.
- Calculate the volume of each disc, and record the results in Table 13.1.
- Calculate and record the percent volume change in each sample relative to the original volume (interior volume of the empty petri dish).

Table 13.1. Shrink-Swell Measurements & Volume Changes

Sample	Radius (cm)	Height (cm)	Volume (cm^3)	% Volume Change
Empty petri dish				N/A
Kaolinite				
Bentonite				

If you were building a house with a basement, would you prefer the dominant clay mineral of the soil on your lot to be kaolinite or bentonite? Explain.

Activity 2. Demonstrations of Colloid Charge

Using the remaining slurry from Activity 1, your instructor will demonstrate colloid charge to the class. In this demonstration, two wires with alligator clips are attached to a 6V battery: one wire to the positive terminal, and the other to the negative terminal. The two wires are then inserted into the slurry and allowed to react for approximately 15 minutes.

Describe what you observe when the wires are removed from the slurry, and explain the phenomenon that causes this.

Activity 3: Demonstration of Flocculation and Dispersion

As you now know, most soil clays have a negative charge that is neutralized to varying degrees by the adsorbed cations. Flocculates or aggregates form most rapidly and are most stable when the soil colloid is most completely neutralized by the adsorbed cation. The amount of neutralization is related to ion size, valence, and concentration. Perform the following experiment to test the effect of various ions on the flocculation of a dispersed soil clay.

1. Fill 5 test tubes to 1/2 full with the dispersed clay suspension. Test tube 5, containing the original clay suspension, will be reserved to use as a check.
2. Add 10 drops of 1M $AlCl_3$ to test tube number 1.
3. Add 10 drops of 1 M KCl to test tube 2.
4. Add 10 drops of 1M NaCl to test tube 3.
5. Add 10 drops of 1M $CaCl_2$ to test tube 4.
6. Shake each suspension thoroughly.
7. Observe and record the time that flocculation starts in each tube.
8. Continue to observe throughout the remainder of the class period and note the relative size and settling rates of the floccules.

Record your observations in Table 13.2.

Table 13.2. Effect of cations on flocculation of a clay suspension.

Added cation	Relative Size & Settling Rates of Floccules
K+	
Na+	
Ca2+	
Al3+	
Check	

Activity 4. Determining CEC by replacing adsorbed cations.

In this activity, you will titrate the filtrate with a 0.01 molar solution of NaOH using phenolphthalein as an indicator. Phenolphthalein changes from colorless to faint pink when the quantity of OH^- ions added via the NaOH equals the quantity of H^+ ions in the solution (that is, when the pH is raised to 7). For this activity, assume the soil samples have been extracted and the filtrates are now available for analysis.

1. Place 10 ml of each filtrate into separate 125 ml flasks. This 10 ml quantity is the amount of filtrate from 1.0 gram of soil.
2. Add 10 drops of the phenolphthalein indicator.
3. Titrate the extract with the NaOH solution to a faint pink endpoint. The titration must be done very carefully to obtain meaningful results. If you put too much NaOH in the flask and get a bright pink color, discard the solution and repeat the process. In the table below, record the milliliters of NaOH solution used to achieve the endpoint.

Calculate the CEC and record your data in Table 13.3.

Here is an example of how to calculate the CEC, assuming 2.5 mL of NaOH was required to achieve an end point. The reaction occurring during titration is

$$\mathrm{NaOH} + \mathrm{H}^+ \rightarrow \mathrm{Na}^+ + \mathrm{H_2O}$$

Thus, one mole of NaOH reacts with one mole of H^+. Therefore, at the phenolphthalein end point, moles of NaOH added = moles of H+ in solution.

The solution of 0.01 molar NaOH contains 1 cmol charge per liter (1 $\mathrm{cmol_c/L}$). Therefore 2.5 mL NaOH contains

$$\mathrm{cmol_c\ of\ NaOH} = 2.5\ \mathrm{mL\ NaOH} \times \frac{1\ \mathrm{L}}{1000\ \mathrm{mL}} \times \frac{0.01\ \mathrm{mol\ NaOH}}{1\ \mathrm{L}} \times \frac{1\ \mathrm{mol_c}}{1\ \mathrm{mol\ NaOH}} \times \frac{100\ \mathrm{cmol_c}}{1\ \mathrm{mol_c}} = 0.0025\ \mathrm{mol_c\ NaOH}$$

Thus, the CEC is

$$\frac{\mathrm{cmol_c}}{\mathrm{kg\ soil}} = \frac{0.0025\ \mathrm{cmol_c}}{1\ \mathrm{g\ soil}} \times \frac{1000\ \mathrm{g\ soil}}{1\ \mathrm{kg\ soil}} = \frac{2.5 cmolc}{\mathrm{kg\ soil}}$$

Table 13.3. CEC calculations.

Extract	Molarity of NaOH	Milliliters NaOH used	Wt. of soil	CEC	% Clay
Norfolk E	0.01		1 g		9
Norfolk Bt	0.01		1 g		37
Cecil Ap	0.01		1 g		13
Cecil Bt	0.01		1 g		51
White Store Bt	0.01		0.5 g		57

Using the data from Table 13.3, construct a graph in Figure 13.2 with percent clay on the X-axis and CEC on the Y-axis. Label the axes and plot all data. Draw a line through the points on the graph for the Norfolk E, Norfolk Bt, and Cecil Bt, and answer the following questions:

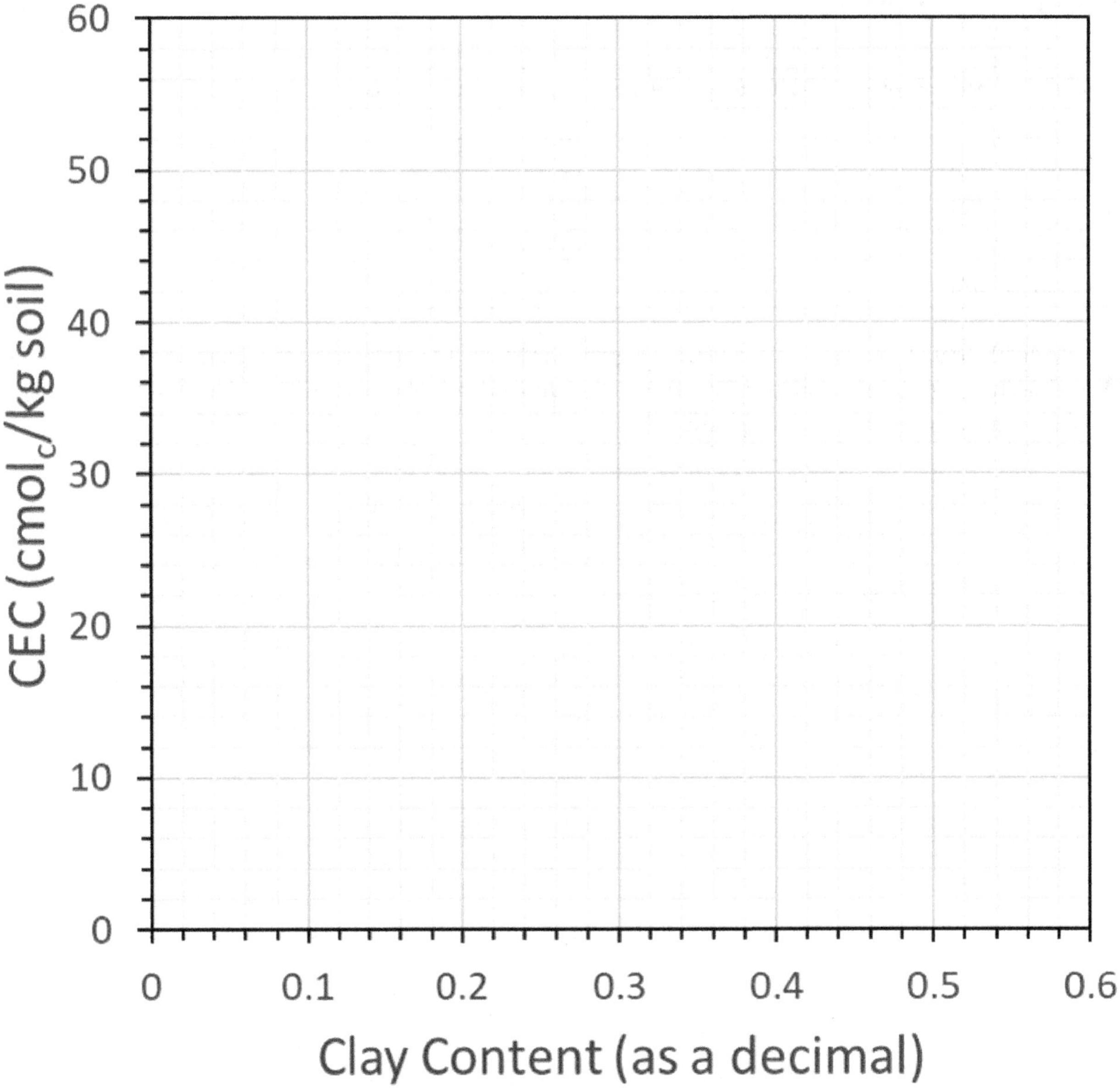

Figure 13.2. Relationship between clay content and CEC.

Of the four major factors that affect soil CEC (amount of clay, type of clay, amount of humus, pH), which is responsible for this linear increase in CEC?

The slope of the line represents the change in CEC divided by the change in clay content. By expressing clay content as a fraction (for example, 30% clay = 0.30 kg clay/kg soil), the slope becomes

$$\text{Slope} = \frac{\Delta \frac{\text{cmol}_c}{\text{kg soil}}}{\Delta \frac{\text{kg clay}}{\text{kg soil}}} = \frac{\text{max CEC} - \text{min CEC}}{\text{max clay} - \text{min clay}} = \_\_\_\_\_\_ \frac{\text{cmol}_c}{\text{kg clay}}$$

The calculated slope is the CEC on a clay basis. Using that slope (cmol_c/kg clay), use Figure 13.2 to determine the type of clay in the Norfolk and Cecil soils.

The line you have drawn does not pass through the origin, indicating that a soil with no clay would still have CEC. How could this be possible?

The Cecil Ap horizon sample falls just above the line you drew in question #1, yet both Cecil samples were taken from the same profile. How can you explain this difference in CEC?

The CEC for the White Store Bt horizon does not fall on the line for the Norfolk and Cecil soil either. Explain what may be different about the colloids in the White Store Bt soil.

Example Calculations of Cation Exchange Capacity

Assume a soil has a CEC of 1.0 cmol_c/kg. How many kg of Ca^{2+} can be adsorbed in a hectare of soil to a depth of 20 cm? Assume a bulk density of 1.4 Mg/m^3.

Soil volume = area depth:

$$\text{Soil volume} = 10,000\ \text{m}^2 \times 0.2\ \text{m} = 2,000\ \text{m}^3$$

Soil weight = density volume:

$$\text{Soil weight} = 1.4 \frac{\text{Mg}}{\text{m}^3} \times 2,000\ \text{m}^3 = 2,800\ \text{Mg} = 2,800,000\ \text{kg}$$

Because Ca has a valence of 2, one mole of Ca has 2 moles of charge:

$$\frac{2\ \text{mol}_c}{1\ \text{mol Ca}|^{2+}} \text{ or in terms of cmols, } \frac{2\ \text{cmol}_c}{1\ \text{cmol Ca}|^{2+}}$$

The soil can adsorb 1.0 cmol_c/kg. The amount of Ca required to supply this amount of charge is

$$\frac{1\ \text{cmol}_c}{\text{kg soil}} \times \frac{1\ \text{cmol Ca}^{2+}}{2\ \text{cmol}_c} = \frac{0.5\ \text{cmol Ca}^{2+}}{\text{kg soil}}$$

The atomic weight of Ca is 40.078, so the soil can adsorb

$$\frac{0.5 \text{ cmol } Ca^{2+}}{\text{kg soil}} \times \frac{1 \text{ mol } Ca^{2+}}{100 \text{ cmol } Ca^{2+}} \times \frac{40.078\text{g } Ca^{2+}}{1\text{mol } Ca^{2+}} = \frac{0.20 \text{ g } Ca^{2+}}{\text{kg soil}}$$

The quantity of Ca adsorbed in the total soil volume is

$$\frac{0.20 \text{ g } Ca^{2+}}{\text{kg soil}} \times \frac{2,800,000 \text{ kg soil}}{1 \text{ ha}} \times \frac{1 \times \text{ kg } Ca^{2+}}{1000 \text{ g } Ca^{2+}} = \frac{560 \text{ kg } Ca^{2+}}{ha}$$

In the problem above, how much Al^{3+} could be adsorbed by the soil?

Because the valence of Al is 3, one mole of Al has 3 moles of charge. Thus, the soil, which has 1 $cmol_c$/kg, can adsorb

$$\frac{1 \text{ cmol}_c}{\text{kg soil}} \times \frac{1\text{cmol } Al^{3+}}{3 \text{ cmol}_c} = \frac{0.33 \text{ cmol } Al^{3+}}{\text{kg soil}}$$

The atomic weight of Al is 26.981538 so the soil can adsorb

$$\frac{0.33 \text{ cmol } Al^{3+}}{\text{kg soil}} \times \frac{1 \text{ mol } Al^{3+}}{100 \text{ cmol } Al^{3+}} \times \frac{26.982 \text{ g } Al^{3+}}{1 \text{ mol } Al^{3+}} = \frac{0.09 \text{ g } Al^{3+}}{\text{kg soil}}$$

The quantity of Al adsorbed in the total soil volume is

$$\frac{0.09 \text{ g } Al^{3+}}{\text{kg soil}} \times \frac{2,800,000 \text{ kg soil}}{1 \text{ ha}} \times \frac{1 \times \text{ kg } Al^{3+}}{1000 \text{ g } Al^{3+}} = \frac{252 \text{ kg } Al^{3+}}{ha}$$

Cation Exchange Capacity Calculations

Calculate how many kg/ha of the following cations this soil (with a CEC of 1 $cmol_c$/kg) could adsorb.

Mg^{2+}, with an atomic weight of 24.305:

Na^{+}, with an atomic weight of 22.990:

H^+, with an atomic weight of 1.008:

K^+, with an atomic weight of 39.098:

Base Saturation Calculations

Ca^{2+}, Mg^{2+}, K^+, and Na^+ are called basic cations because when they are in solution, the solution is basic. In contrast, H^+ and Al^{3+} are considered acidic cations because they lower solution pH.

Therefore, base saturation is

$$\text{Percent base saturation} = \frac{\text{cmol}_c \text{ of base cations}}{\text{total CEC of the soil}} \times 100\%$$

An alkaline soil found in a semi-arid region might have the following cation exchange characteristics (per kg of soil):

- 9.0 cmol_c Ca^{2+}
- 2.5 cmol_c Mg^{2+}
- 0.5 cmol_c K^+
- 2.0 cmol_c Al^{3+}
- 2.0 cmol_c H^+

If these cations are the only ones on the exchange sites of this soil, what is the percent base saturation?

Activity 5. Calculating versus estimating CEC

There are two ways you can calculate the CEC: the sum of cations method and the mineralogy method.

The Sum-of-Cations Method

If you have a soil analysis where the quantities of all cations in the soil are listed, simply summing all those exchangeable quantities will yield the CEC you found in the preceding problems.

The "Mineralogy" Method

As you know from your reading and class discussion, clay minerals have a range of values for CEC. If the mineralogy of the clay fraction is known (that is, the type and amounts of each clay mineral), then the CEC can be approximated.

To make these calculations easier, Table 13.4 contains representative values for CEC to use in all calculations for this class unless otherwise noted. In nature, however, these soil colloids will have a range of values.

Table 13.4. Typical CEC of various soil colloids.

Mineral or colloid type	CEC of pure colloid
	cmolc/kg
kaolinite	10
illite	30
montmorillonite/smectite	100
vermiculite	150
humus	200

As an example of this mineralogy approach to CEC calculations, consider a soil having 100% clay where the clay is 100% kaolinite. The CEC would then be 10 $cmol_c/kg$. If a soil contains only 10% kaolinite (or 10 kg clay in 100 kg soil), however, this clay would contribute

$$\text{Total CEC of the soil} = \frac{10\ \text{cmol}_c}{\text{kg clay}} \times \frac{10\ \text{kg clay}}{100\ \text{kg soil}} = \frac{1.0\ \text{cmol}_c}{\text{kg soil}}$$

A prairie soil contains 30% clay. This clay sized fraction is dominantly montmorillonite. The soil also contains 5% humus (organic matter).

Using the mineralogy method, what is the cation exchange capacity (CEC) contributed by the clay?

What is the estimated cation exchange capacity (CEC) contributed by the humus?

What is the total estimated CEC of this soil?

The following is actual laboratory data for the soil in the example:

Table 13.5. Exchangeable cation data

Exchangeable Acidity	Exchangeable Bases	Exchangeable Bases	Exchangeable Bases	Exchangeable Bases
H+ + Al3+	Ca2+	Mg2+	K+	Na+
----cmolc/kg----	----cmolc/kg----	----cmolc/kg----	----cmolc/kg----	----cmolc/kg----
14.0	29	10	5.5	1.5

Calculate the CEC from the data in the table using the sum of cations method.

What is the percent base saturation?

If you wanted to replace the Na^+ with Ca^{2+}, how many kilograms of Ca^{2+} per 2,800,000 kg of soil would you need?

Assignment: Online Quiz

A quiz for this lab will be available online. Please access it as directed by your instructor.

14. Soil Acidity and Adjusting Soil pH

Managing soil pH is essential to creating ideal growth conditions for most plants. This is because the pH of the soil controls the solubility of nutrients as well as toxic metals. Because of this, most plants have a preferred range in soil pH. Under most cases, liming agents are added to soil to raise pH to the desired range. However, in some cases, lower soil pH is desired, which can be achieved using soil amendments such as elemental sulfur (S), or aluminum sulfate (more commonly referred to as "alum"). In either case, the pH tolerance of the target plant species, the properties of the soil, and the properties of the soil amendment must be considered to achieve the desired change in soil pH.

Learning Objectives

- Understand the origin of soil acidity.
- Measure soil pH with field and laboratory techniques.
- Determine the role of aluminum in soil acidity.
- State the relationship between cation exchange capacity, buffering capacity, and potential acidity.
- Write chemical reactions related to soil pH and liming.
- Know the objectives of liming and the factors affecting lime requirement.
- Measure limestone requirement.
- Determine limestone quality.

Materials

- Soils with low pH
- pH meter with a pH electrode
- pH test strips
- Beakers
- Glass stir rods
- Pure reagent grade calcium carbonate
- Pure reagent grade calcium oxide
- Pure reagent grade calcium sulfate (gypsum)
- Dolomitic limestone, coarse (sieved using a 20-40 mesh sieve)
- Dolomitic limestone, fine (sieved using a 100+ mesh sieve)

Recommended Reading & Viewing

- Soil pH Overview (CropWatch – Youth, 2013c)
- Soil pH Test (CropWatch – Youth, 2013d)
- Soil pH (USDA-NRCS, 2011)
- Soil pH (USDA-NRCS, 2014a)
- Liming Acid Soils (Whitney and Lamond, 1993)

Prelab Assignment

Using the recommended reading and the introduction to this lab, consider the questions listed below. These definitions/questions will provide a concise summary of the major concepts addressed in the lab. They are also useful as study notes for exams.

1. Define pH and pOH. Show the formulas used to calculate both.
2. Why is aluminum considered an acidic cation? Show the relevant reactions.
3. List four negative consequences of low pH (acid) conditions in soil. List two negative consequences of high pH (basic) conditions in soil.
4. Define buffering capacity. How does buffering capacity relate to cation exchange capacity?
5. Define active acidity, salt-replaceable (exchangeable) acidity, and residual acidity.
6. List various soil amendments that are used to increase the pH of an acid soil?
7. Write a chemical reaction depicting how a typical liming material (calcitic limestone, burned lime, etc.) act to neutralize acidity.
8. List various soil amendments used to acidify a soil and reduce the pH.
9. What influences how much material is needed to increase the pH of an acid soil?

Introduction

Soil acidity is largely controlled by the composition of ions on exchange sites on the colloidal fraction. The H^+ cations are acidic by definition, and Al^{3+} cations are considered acidic because they react with H_2O to produce $Al(OH)_3$ and $3H^+$. The Ca^{2+}, Mg^{2+}, K^+, Na^+ cations, among others, are considered basic because they form strongly dissociated bases by reacting with OH^-. These exchangeable cations on the exchange complex are in equilibrium with the cations in soil solution. Therefore, the nature of the exchangeable cations influences the composition of the soil solution.

The acidic cations adsorbed on the negative exchange sites are called the *reserve* (also *residual* or *potential*) and *salt-replaceable* (also *exchangeable*) acidity. The reserve and salt-replaceable acidity controls the level of soluble or *active* acidity in the soil solution. Only the active acidity is measured in a routine pH determination. The reserve and salt-replaceable acidity is always many times higher than the active acidity.

A soil is acid when hydrogen ions predominate in the soil. The degree of acidity is expressed in terms of pH, which is defined as the negative logarithm of the hydrogen ion activity. Therefore, the pH of a 0.01-molar hydrogen ion solution is

$$\mathrm{pH} = -\log\left(\frac{10^{-2}\ \mathrm{mol\ H^+}}{\mathrm{L}}\right) = 2$$

At pH 7, the concentration of H+ ions and OH- ions are equal, and the soil or solution is neutral. At pH values less than 7, the soil is acid; at values more than 7, the soil is alkaline. Most soils vary in pH from about 4 to 10. Soils in areas with high rainfall are generally acid with a pH less than 7. Soils developed in high-lime deposits often will be alkaline. Soils high in calcium seldom have pH values higher than 7.5, but the presence of large amounts of calcium carbonate may cause the pH to be as high as 8.5. Where the pH is higher than 8.5, an excess of sodium is highly probable.

The most desirable soil pH for most crops in Kansas is 6.8. However, crops like blueberries need a lower pH, and other crops, like alfalfa, need a higher pH. At soil pH less than 5.8, several problems may occur:

- Al and Mn toxicity
- Inhibited growth of N-fixing bacteria
- Possible deficiencies in Mg and/or Ca.
- P deficiency (P reacts with Fe and Al)
- At more than pH 7.5, other problems may occur:
- Deficiency of Fe, Mn, Cu, or Zn
- P deficiency (P reacts with Ca)

Buffering Capacity

Buffering capacity is a measure of the soil's ability to resist a change in pH, directly related to the magnitude of the exchange capacity. Small fluctuations in acid or base content can occur without a noticeable pH change as cations are adsorbed or released from the exchange complex. Soils with the largest cation exchange capacity have the greatest buffering of a pH change. In other words, two soils may have the same pH (active acidity in soil solution), but the one with the largest cation exchange capacity will have the most acidity stored in reserve and therefore the highest buffering capacity or ability to resist a change in pH. For this reason, it takes less lime to increase the pH of a sandy soil (low CEC) by a given amount than it takes to increase the pH of a clay soil (higher CEC) the same amount.

Sources of Soil Acidity

Controlling soil pH is vital to optimal use and productivity of soils. Adding lime is the most effective and practical way to raise the pH of acid soils. Elemental sulfur, iron sulfate, or aluminum sulfate can be used to reduce soil pH. Because acidity is a concern in Kansas, we will focus on raising soil pH. Understanding the following equations should help you understand the sources of soil acidity and soil reactions to lime.

Acid cations adsorbed to colloids can be released to solution through cation exchange, in which cations such as Ca^{2+}, Mg^{2+}, K^+, etc. can displace H^+ and Al^{3+}, forcing them into the soil solution, moving those acid cations from the salt-replaceable acidity pool to the active acidity pool. Notice that the reaction is reversible, so having high amounts of acid cations in solution could also cause the displacement of base cations from colloid exchange sites.

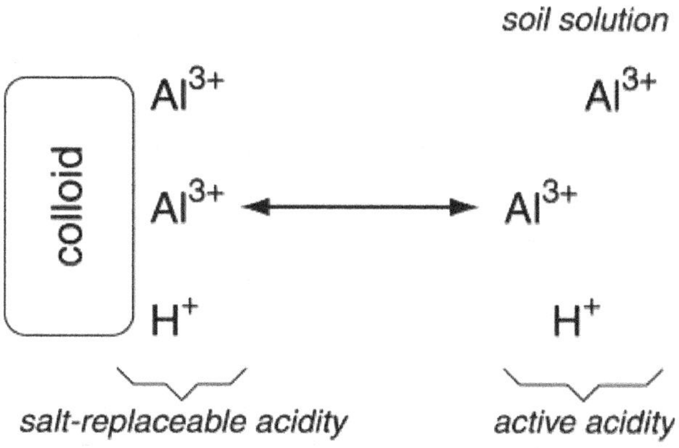

One product of respiration is CO_2. In the soil, the respiration of bacteria, fungi, protists, roots, etc. contributes to a very high concentration of CO_2 in the soil air. When CO_2 becomes dissolved in the soil solution, it reacts with water to form carbonic acid – a weak acid that can release H^+ into solution, thus lowering the soil pH. Note that this reaction is reversible.

$$H_2O + CO_2 \leftrightarrow H_2CO_3 \leftrightarrow H^+ + HCO_3^-$$

Nitrification occurs under aerobic conditions. The oxidation of NH_4^+ to a final product of NO_3^- is facilitated by nitrosomonas and nitrobacter bacteria. The net reaction is shown below. Note that this reaction is not reversible, and that for every two moles of NH_4^+, there is four moles of H^+ released.

$$2NH_4^+ + 4O_2 \rightarrow 2NO_3^- + 2H_2O + 4H^+$$

Sulfur can be used as a soil amendment for lowering soil pH. A common example is for growing plants that prefer acidic conditions, such as blueberry, in soils that are neutral to alkaline. In this reaction, elemental sulfur is oxidized to form sulfuric acid – a strong acid. This is a microbial mediated process.

$$2S + 3O_2 + 2H_2O \rightarrow H_2SO_4^+ \leftrightarrow SO_4^{2-} + 2H^+$$

Aluminum is considered an acidic cation due to the release of H^+ during aluminum hydrolysis. For every one mole of Al^{3+}, three moles of H^+ are released into solution. Note that this reaction is reversible.

$$Al^{3+} + H_2O \leftrightarrow Al(OH)^{2+} + H^+$$

$$Al(OH)^{2+} + H_2O \leftrightarrow Al(OH)_2^+ + H^+$$

$$Al(OH)_2^+ + H_2O \leftrightarrow Al(OH)_3 + H^+$$

Raising soil pH with lime

Standard agricultural lime, which is primarily calcium carbonate ($CaCO_3$) is added to soils to increase soil pH. The $CaCO_3$ reacts with water as shown below.

$$CaCO_3 + H_2O \rightarrow Ca^{2+} + HCO_3^- + OH^-$$

The Ca^{2+}can displace other cations on the cation exchange, including H+ and Al^{3+} (salt-exchangeable acidity), thus releasing it into solution. However the Ca^{2+} does not react (neutralize) the acid cations. The acid cations are neutralized by HCO_3^- and OH^-. An example of the overall chemical reaction following the addition of lime to acid soils is represented below.

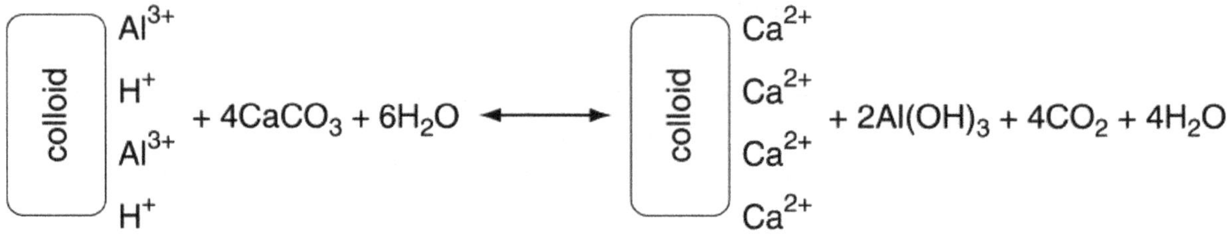

Chemical reaction following the addition of lime.

There are liming agents other than $CaCO_3$. Because most of those other liming agents have different chemical formulas, the chemical reactions are different. This also means that the effectiveness of different liming agents vary from that of $CaCO_3$.

Factors that Affect Liming Rates

The major factors that impact lime rates (the amount of lime required to raise the soil pH to a target pH) include the crop requirement; the type, size, and purity of the liming agent; the cation exchange capacity of the soil; and the pH of the soil.

Most plants have a range in soil pH in which they are most health, or produce the highest yield. This is because soil pH impacts nutrient availability. For example, some plants, like alfalfa and sweet clover, require more calcium than others, and thus require higher rates of liming. Others, like azaleas, cranberries, and blueberries, require more iron; which is more soluble at a lower pH. Therefore, it would take a larger application rate of lime to raise the soil to the desired soil pH for alfalfa than it would for blueberries.

It takes time for limestone to dissolve and replace hydrogen or aluminum on the soil exchange positions. Smaller lime particles have a larger surface area that is exposed and available to react, which reduces the time required for all of the lime to react. Therefore, finer liming agents are more effective at raising soil pH. Many limestones are predominately calcium carbonate ($CaCO_3$), and some contain both $CaCO_3$ and $MgCO_3$. However most limestones contain some amount of impurities and inert material. Thus, a liming agent with lots of impurities (and less lime) is less effective at raising soil pH. Obviously, the purer the $CaCO_3$, the more effective the lime is.

Soils with the same pH may require different amounts of limestone due to differences in CEC, which would imply differences in buffering capacities. For example, consider the amount of limestone necessary to raise the base saturation of two soils from 70% to 90% when one soil has a CEC of 15 cmol_c/kg, and the other has a CEC of 40 cmol_c/kg.

$$15\frac{\text{cmol}_c}{\text{kg}} \times 20\% \text{ increase} = 3\frac{\text{cmol}_c}{\text{kg}} \text{ basic cations required from lime}$$

$$40\frac{\text{cmol}_c}{\text{kg}} \times 20\% \text{ increase} = 8\frac{\text{cmol}_c}{\text{kg}} \text{ basic cations required from lime}$$

Lastly, soil pH is governed by base saturation. If other factors are constant, the lower the pH, the more lime that is required to achieve a desired pH. This is because at a low pH, a larger percentage of the CEC is occupied by acid cations, which requires larger amounts of lime to neutralize.

Activity 1: Determining pH With Indicator Strips (Field Method)

Of the several techniques available for determining pH, one that can be used easily in the field is the indicator strip method. This technique uses the principle of pH sensitivity of certain dyes, which cause differences in color across a range in pH. With the soils provided, complete the following pH determination:

Weigh 10.0 g of soil into a small plastic cup. Add 20 ml of distilled water and stir. Allow to stand for 5 minutes, occasionally stirring.

Using the pH indicator strips provided, dip the strip into the cup until the tip is wetted. Determine the pH by comparing the color change of the pH test strip to the color chart.

Record the soil pH in Table 14.1.

Activity 2: Determining Soil pH with a pH Meter

Laboratory pH meters are more accurate than pH dyes and strips. The pH meter measures the hydrogen ion activity $[H^+]$ by measuring the electric potential across a thin, porous glass membrane at the base of the electrode. This potential changes in response to $[H^+]$, and by standardizing the instrument with buffers of known pH, we can measure the pH of any solution, including soil solutions.

Using the samples prepared in Activity 1, carefully place the electrode in the suspension. Gently swirl the electrode in the solution, and note the pH reading. Wait for the pH meter to reach a steady reading, indicated by the word "ready" on the screen.

Record the value for this 1:2 soil-water suspension in Table 14.1.

Activity 3: Determining a Need for Adding Limestone to the soil

To decide if a soil needs lime, you need to know only the pH, or the active acidity, of the soil solution. If the pH is less than or equal to 5.8, lime is generally recommended (the pH below which lime is required varies by region and intended crops).

For each soil analyzed in Activity 2, decide if limestone is needed and record your decision in Table 14.1.

Table 14.1. Results for Activities 1-4: Determining Soil pH and Limestone Need

Soil	Soil pH (strip)	Soil pH (meter)	Lime needed?	Lime Requirement
			Yes/No	(lbs ECC/ac)
A				
B				
C				
D				

Activity 4: Determining How Much lime is needed

To decide how much lime a soil needs, you must determine the amount of reserve acidity in the soil. This reserve acidity is often called exchangeable acidity because it can be dissociated from the cation exchange complex through which it enters the soil solution. To determine the exchangeable acidity, a buffer solution of known pH is added to the soil. This buffer solution contains cations that will replace H^+ and Al^{3+} on the exchange complex. The acidic cations removed from the exchange complex reduces the pH of the added buffer. Figure 14.1 illustrates the decrease in pH of a buffer solution as acid is added. Once a relationship is determined (shown by the slope of the line), you can add soil to the buffer solution (or buffer solution to the soil) to determine the exchangeable acidity.

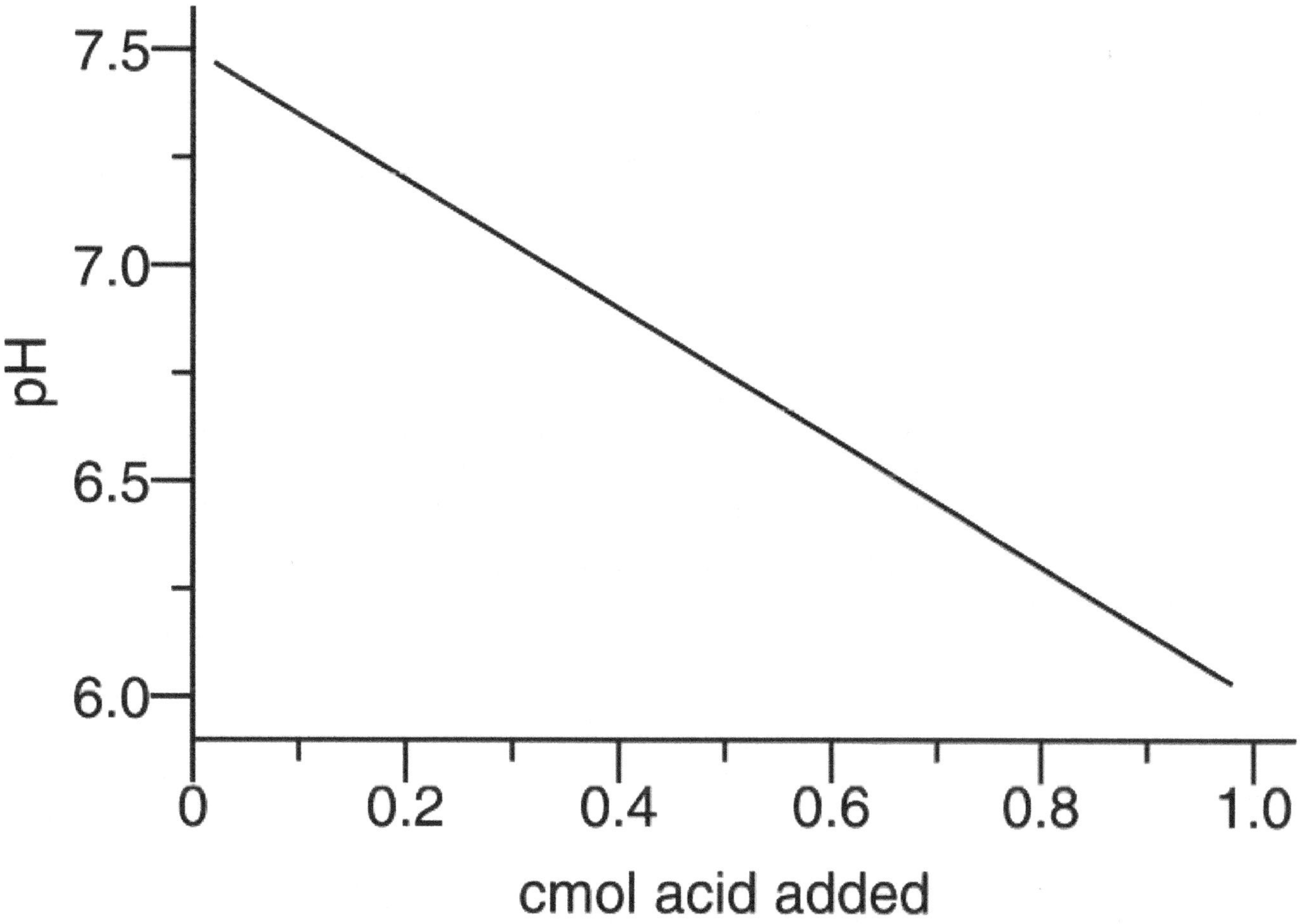

Figure 14.1. Change in pH of a buffer solution as acid is added. Diagram courtesy of King et al. (2003).

In this exercise, we will use the SMP buffer – the buffer solution used by the K-State Soil Testing Lab. This buffer solution was designed to provide the liming rate when using the following formulas, depending on the region and target pH (see bullets).

$$\text{Target pH of 6.8} =$$

$$[25,620 - (6,360 \times \text{ buffer pH}) + (391 \times \text{buffer pH} \times \text{buffer pH})] \times \text{depth}$$

- Depth is in inches
- Used for all crops in Southeast Kansas (east of Flint Hills and south of Highway 56)
- Used for alfalfa and clover in Northeast Kansas
- Lime is recommended if pH < 6.4

$$\text{Target pH of 6.0} =$$

$$[12,810 - (3,180 \times \text{ buffer pH}) + (196 \times \text{buffer pH} \times \text{buffer pH})] \times \text{depth}$$

- Depth is in inches
- Used for all crops in Northeast Kansas other than alfalfa and clover
- Used for all crops in Central and Western Kansas

- Lime is recommended if pH < 5.8

$$\text{Target pH of 5.5} =$$

$$[6{,}405 - (1{,}590 \times \text{ buffer pH}) + (98 \times \text{buffer pH} \times \text{buffer pH})] \times \text{depth}$$

- Depth is in inches
- Used if cash flow is limited or in lime availability problem areas in Central and Western Kansas
- Lime is recommended if pH < 5.5

This buffer contains chromium (Cr), a toxic heavy metal. Therefore, your lab instructor will perform the SMP buffer analysis. As a class, determine which soil-water mixtures from Activity 1 need lime (pH ≤ 6.4). To those solutions, add 10 ml of the SMP buffer solution, and stir with a glass rod. Allow the mixtures to stand for 30 minutes, which should be enough time for the acid cations to be displaced from the CEC and forced into solution. Read the pH on meter.

Assuming the desired pH is 6.0 (i.e. use the middle equation), calculate the lime requirement, show your work below, and record your results in Table 14.1.

Activity 5: Evaluating Liming Materials

The type of liming material and the size or fineness of the material determine how efficiently liming materials raise soil pH. This experiment was actually initiated earlier in the semester to allow time for the liming agents to react. Amending the soil with several different liming agents allows us assess the effects of particle size and liming material based on the relative changes in soil. The treatments included the following:

- Reagent grade $CaCO_3$
- Reagent grade CaO
- Reagent grade $CaSO_4$
- Coarse dolomitic limestone (35 mesh)
- Fine dolomitic limestone (120 mesh)
- Control (no amendments)

When this experiment was initiated, each lab section was divided into six groups, with each group responsible for one of the six treatments. Your laboratory instructor assigned a treatment to your group, and you completed the following steps:

1. Label four plastic bags
2. Weigh 20 g of air-dry soil into each plastic bag.
3. Weigh 0.1 gram of designated liming material onto weighing paper.
4. Add the liming material to the soil and mix thoroughly to distribute evenly in the soil.
5. Add a few mL of water to each bag and mix.
6. Close the bags to start incubation.

Now that the liming agents have had time to react, you will collect the results.

7. Add 40 ml of distilled water to the plastic bag and mix. Let it sit for five minutes, mixing occasionally.
8. Carefully place the electrode in the suspension, swirl the electrode, and determine the pH like you did previously in Activity 2.
9. Record the pH for this 2:1 water-soil solution in Table 14.2 below.

Table 14.2. Liming material experiment results

Treatment	Soil 1 (Sandy Soil) pH	Soil 2 (Clayey Soil) pH
Pure reagent grade $CaCO_3$		
Pure reagent grade CaO		
Pure reagent grade $CaSO_4$		
Dolomitic limestone (35 mesh)		
Dolomitic limestone (120 mesh)		
Control (no amendments)		

Record your group's data on the table on the blackboard. Then record all of the class data onto the table above.

Using that data, we can examine the magnitude of pH change for the different liming agents and the two different textures. For each of the data points, subtract control pH from the resulting pH to determine the magnitude of pH change for each liming agent. Record those results in Table 14.3. Also, calculate the average pH change for each liming agent across both soil types, and the average pH change for each soil type across all liming agents. Record those values in Table 14.3 also.

Treatment	Clayey pH	Sandy pH
Pure Reagent-grade $CaCO_3$		
Pure Reagent-grade CaO		
Pure Reagent-grade $CaSO_4$		
Dolomitic Limestone (20-40 mesh)		
Dolomitic Limestone (120 mesh)		

As a class, graph the average change in pH for each liming material in a bar chart, and do the same for the average pH change for both soil textures. Then discuss the results as a class.

Activity 6: How Soil Characteristics affect Liming Reaction

To illustrate the effects of soil texture and organic matter on pH adjustment, your lab instructor added various rates of dolomitic limestone to a sandy soil, a clay soil, and an organic soil. The rates of limestone ranged from zero to 10 tons/acre. After several months, the following pH values were measured:

Table 14.4. Resulting pH values following lime additions

Tons of added lime	**Resulting pH Values**		
	Sandy soil	Clay soil	Organic soil
0	4.5	4.5	3.5
1	5.5	4.9	3.6
2	6.5	5.2	3.7
3	7.0	5.5	3.8
4	7.2	5.7	3.9
5	7.3	5.9	4.0
6	7.4	6.1	4.2
7	7.5	6.3	4.5
8	7.6	6.5	5.0
9	7.7	6.7	5.5
10	7.7	7.0	6.0

Plot the results in Table 14.3 into Figure 14.2 for all three soils, and connect the data points to form a line for each of the three soils. . Using the following graph to plot the results of pH against tons of lime added for each of the soils, answer the questions about this graph.

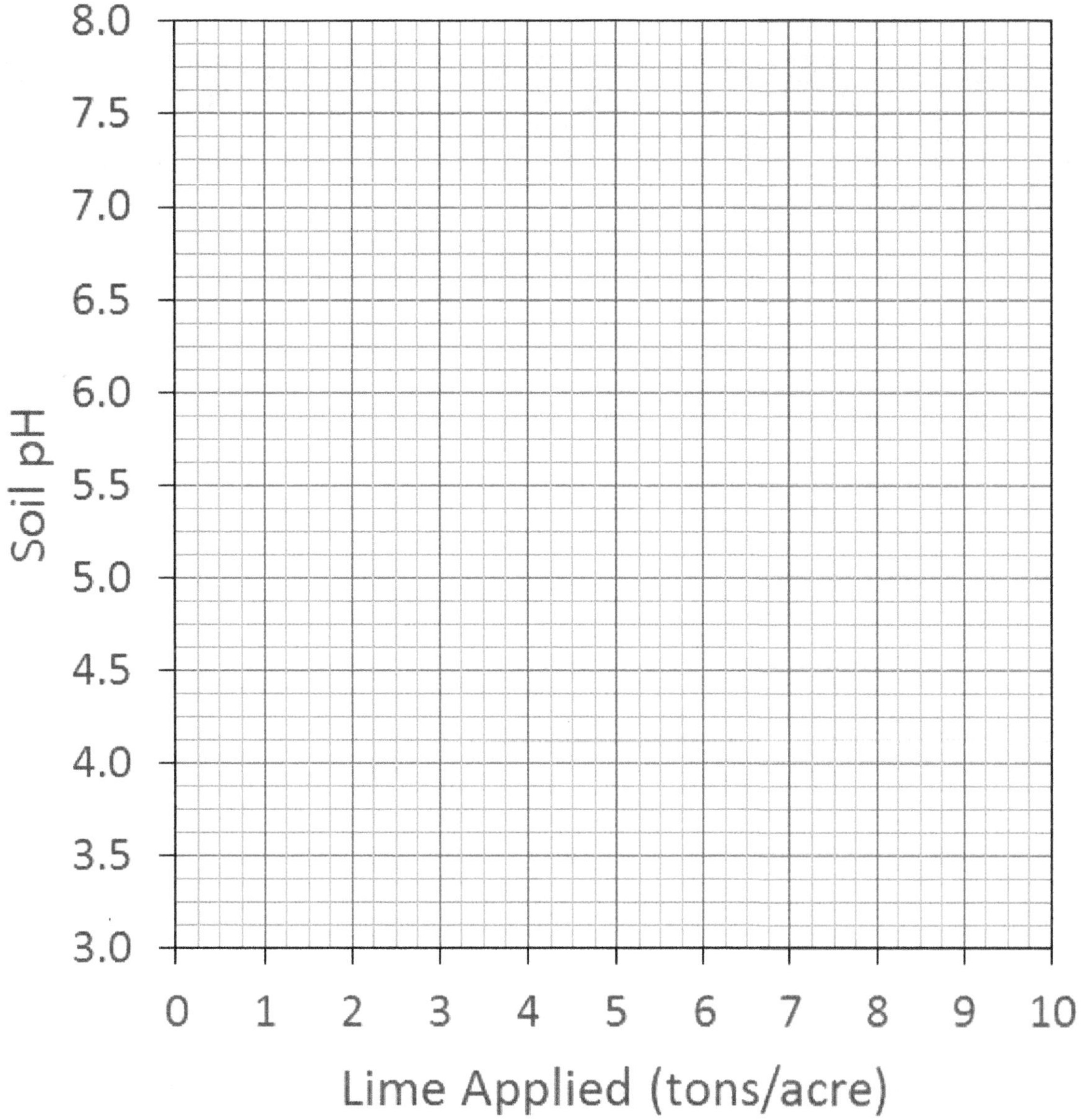

Figure 14.2. Soil pH response to liming rates for three soils.

What is the relationship between the amount of lime necessary to raise soil pH and the cation exchange capacity of the soil?

Should the pH goals for a clay soil and an organic soil always be the same? **Explain** your answer. HINT: Organic soils contain very little Al.

If the sandy and the clayey soil were limed to achieve the same pH, which soil will probably need lime sooner? Why?

Assignment: Online Quiz

A quiz for this lab will be available online. Please access it as directed by your instructor.

PART VI

SOIL FERTILITY AND MANAGEMENT

15. Soil Nutrient Management

Proper soil fertility is essential for achieving optimum yields and plant health. It is also critical for minimizing the costs of inputs like fertilizer or irrigation water, as well as for reducing the environmental impacts of production agriculture or manicured landscapes. The key to proper soil fertility is soil testing. We will first tour the K-State Soil Testing Lab. The staff will walk the class through the process, include sample submittal, costs, pH analysis, and nutrient analyses. This will be followed by a problem set in which students will make fertilizer recommendations using soil test results and publically available K-State extension bulletins and resources.

Learning Objectives

- Learn to interpret information contained in a soil test report.
- Make fertilizer recommendations based on soil test results.
- Become familiar with K-State resources as well as publications available to farmers and gardeners.

Materials

- Soil Nutrient Management Problem Set
- Fertilizing Kansas Lawns (Fagerness, 2001)
- Fertilizer Recommendation Program (Kansas State University Soil Testing Laboratory, Kansas State University Department of Agronomy, Manhattan, KS, U.S.)
- Soil Test Interpretations and Fertilizer Recommendations (Leikam et al., 2003)
- Fertilizing Gardens in Kansas (Upham and Marr, 2011)

Recommended Reading

- Soil Analysis (Kansas State University Department of Agronomy, 2020)

Prelab Assignment

Using the recommended reading and the introduction to this lab, consider the questions listed below. These definitions/questions will provide a concise summary of the major concepts to be addressed in the lab. They are also useful as study notes for exams.

1. Explain why it is important to collect a soil sample from a representative area of soil and why it is common practice to mix many samples into one composite sample.
2. Explain the statement that "soil testing laboratories do not determine the concentration of plant-available nutrients in soils". What do they actually determine?
3. Explain why applying fertilizer to achieve maximum yield may not achieve maximum profit.
4. In addition to soil testing, what other diagnostic tools are used to assess fertility?

Activity & Assignment: Problem Set

A problem set will be provided to you at the beginning of the lab. It will familiarize you with how fertilizer recommendations are made based on soil tests. Materials required to complete the problem set are available as printed copies in the lab or as PDFs and Excel Spreadsheets online. You will complete the problem set during lab.

16. Soil Erosion and Conservation

It takes up to 2,000 years to form one inch of soil. Under the right conditions, that inch of soil can be eroded by wind or water in a single storm. Thus, if not conserved, soil that developed over millennia can be lost within a single human generation. However, soils can be used over and over again using appropriate conservation practices. In this lab, several videos will be used to demonstrate the mechanics underlying soil erosion. In addition, a model will be used to estimate the impact of common soil conservation practices on soil erosion.

Learning Objectives

- Identify conditions under which soil is most susceptible to erosion by water.
- Identify conditions under which soil is most susceptible to erosion by water.
- Utilize the RUSLE equation to estimate the erosion rates of a given hillslope.
- Compare estimated erosion rates to "tolerable" rates of erosion, commonly known as T values.
- Name and describe conservation practices that reduce water and wind erosion.
- Become familiar with the federal, state, and local agencies associated with soil and water conservation.

Materials

- Computer with internet access and a projector

Note

For this lab you will need to bring your textbook for lecture, and a laptop or tablet. If you do not have a laptop or tablet, please share with a partner.

Recommended Reading

- Module 3 – Managing for Soil Erosion (McCauley and Jones, 2005a)
- Soil Quality Resource Concerns: Soil Erosion (USDA NRCS, 1998)

Prelab Assignment

Using the recommended reading and the introduction to this lab, consider the questions listed below. These definitions/questions will provide a concise summary of the major concepts to be addressed in the lab. They will also serve as the basis for the post-lab quiz and are useful study notes for exams.

1. Describe the difference between geologic erosion rates, and erosion rates exhibited by managed soils.
2. Identify the soil particle size most susceptible to erosion by wind.
3. Identify the soil texture class size is most susceptible to erosion by water.
4. Identify and describe the three methods of sediment transport by wind.
5. Define sheet, rill, and gully erosion by water.
6. Name the factors included in RUSLE.
7. Identify and describe three conservation practices used to reduce water erosion
8. Identify and describe two conservation practices used to reduce wind erosion.

Introduction

Worldwide there are approximately five billion hectares of degraded land, which is approximately 43% of the Earth's land surface. Of those five billion hectares, the most common cause of degraded land (two billion hectares) is soil degradation due to erosion, compaction, surface crusting, acidification, or salt accumulation. Most of that degradation (85%) comes from water and wind erosion.

However, erosion is a natural process. Geologic erosion is a process that transforms soil into sediment. It takes place naturally without the influence of any human activities. Geologic erosion rates are what has carved and shaped all of the landscapes that we inhabit today. Typical geologic erosion rates are approximately 1 ton/ac annually.

Accelerated erosion occurs when human activities increase the rates of erosion to well above the rates of geologic erosion. It occurs when people disturb the soil, or the vegetation covering the soil. Such practices include overgrazing livestock, cutting forests, plowing hillsides, or tearing up land for construction projects. Accelerated erosion can be 10 to 1000 times as destructive as geologic erosion.

Soil erosion includes two separate processes – soil erosion by water, and soil erosion by wind. Water erosion begins with detachment as rain drops bombard soil aggregates, separating some of them from the aggregate. These stand-alone soil particles are much smaller, and are more easily transported. The transported particles are eventually deposited in a low-lying area, completing the three part process of detachment-transport-deposition. Transport can happen due to splashes from the raindrop, or from running water carrying sediment downhill.

Water erosion begins with sheet erosion where splashed soil is moved uniformly, but some columns of soil that were protected by pebbles may remain. When the water gathers into small channels due to irregularities in the landscape, those channels incise into the soil surface forming a rill. Rills can be smoothed by tillage equipment. If enough water gathers, a gully can form, which is essentially a large rill that is so deep that it cannot be smoothed by tillage equipment. Interill erosion is sheet erosion that occurs between rills. The majority of soil erosion is due to sheet and rill erosion.

Wind erosion is greatest in arid and semiarid regions, such as Kansas, though it can occur to some extent in humid regions. Similarly to water erosion, wind erosion involves three processes – detachment, transportation, and deposition. Detachment occurs as heavy winds push and bounce heavy particles along the surface. As this happens, silt and clay particles can be broken away from aggregates, and can become airborne and transported for great distances. The sediment is transported by three methods – saltation, soil creep, and suspension. Saltation occurs when soil particles move by short bounces, and happens with medium sized particles of 0.1 to 0.5 mm diameter. Soil creep occurs with larger particles, >1.0 mm in diameter, and involves these large particles rolling along the surface. Suspension occurs when particles are suspended in the air for several meters to many kilometers. Only the smallest particles (<0.1 mm diameter) are transported by suspension.

Activity 1: Soil Erosion Videos

Watch the following videos posted to K-State Online, answer the following questions.

Video: Raindrop Impact on a Sandy Surface (Cheng et al., 2014)

Of the three rain drop impacts shown, which velocity seemed to have the largest impact, 1.0 m/s, 3.3 m/s, or 5.4 m/s? (They are shown in that order on the video)

How might the velocity of the raindrop impact detachment? Explain.

How might the velocity of the raindrop impact the distance that soil particles are moved due to splash? Explain.

Video: Rainfall Slow Motion HD Heavy Rain Drops Falling in Slow Mo Video of Droplets Hitting Water (Travel Links Directory, 2013)

Turbulent (violent) flow of water is more erosive to soil particles than laminar (smooth) flow. In this video, is the water turbulent? Also, how might the erosivity of a rain storm change depending on how intense the rainfall is?

Video: Wind Erosion (officemmdivide, 2011)

What is the speed at which raindrops can "pound the ground"?

If the speed of the wind increases from 20 mph to 30 mph, how much does the rate of erosion increase?

Fill in the percentages of sediment transport by each of the three modes of transport by wind:

Table 16.1. Percentages of Sediment Transported by Each Mode of Transport

Mode of Transport	Percentage
Saltation	
Suspension	
Creep	

Draw a diagram below that shows the three types of sediment transport by wind.

Activity 2: Estimating Soil Erosion by Water

In order to estimate current water erosion rates, and to prescribe best management practices (BMPs) that reduce erosion rates, an accurate method to estimate soil erosion was needed. The first model that was used to estimate erosion is known as the Universal Soil Loss Equation (USLE). It is an empirical model, meaning that it was developed from experimentation. It was later revised to form the Revised Universal Soil Loss Equation (RUSLE). For both models, the following equation was used:

$$A = R \times K \times LS \times C \times P$$

Where A is the annual soil loss (tons/ac), R is the rainfall erosivity, K is the soil erodibility, LS is the combined factor of the length of slope and the slope gradient, C is for cover, and P is erosion control factors. The R and K factors cannot be changed through management. The LS factor can be changed through terracing. The C factor can be changed with crop residue management, such as through conservation tillage or cover crops. The P factor can be changed through the use of strip-cropping, contour tillage, or terracing.

The information required for this calculation can be found from a variety of sources. The R factor can be found on maps made available from the Natural Resources Conservation Service (NRCS) or through the extension service. The K factor can be found in the Web Soil Survey. The LS factor is determined from a table of factors (assuming that you have collected the length of the slope and the slope gradient from your field site) or from a topo map. The C factor can be estimated using tables, often from the extension service. Lastly, P can be found from tables provided by the NRCS.

You are going to estimate the erosion rate for slope on the student learning farm, located at the Agronomy North Farm north of Bill Snyder Family Stadium, in Manhattan, KS, which is in Riley County. The erosivity factor (R) for Riley County is 175.

You will find the K value using the Web Soil Survey. Using your computer, navigate to the Web Soil Survey. Navigate to Riley County, Kansas, then zoom in until the field north of the stadium fills most of your screen. The field is bordered by Marlatt Ave to the north, Denison Ave to the east, Kimball Ave to the south, and College Ave to the west. Using the rectangular AOI tool, draw rectangle over the field with those roads as the boundaries, then click on the "Soil Map" tab. Note that the 3919 Smolan silt loam, 1 to 3% slopes is the most prominent soil mapping unit, covering approximately 35% of the field (or approximately 150 ac). Click on the "Soil Data Explorer" tab, then click on the "Soil Properties and Qualities" sub-tab. Expand "Soil Erosion Factors, and click on "K Factor, Whole Soil". Click on "View Rating" to load that data onto the map. Scroll down, and note the "Rating" for the 3919 Map Unit Symbol. Record it below.

Next you will determine the topographic (LS) factor. Navigate back to the "Soil Map" tab. You are going to estimate the erosion rate for the top terrace of the field along College Ave. Your instructor will show you where, precisely. Determine the length of the slope using the "Measure Distance" tool at the top of the map that looks like a lime green ruler. Click at the top of the hill near the ditch, then double-click at the first terrace. Scroll down to the bottom of the screen to determine the segment length. The percent slope is also needed. Typically this is measured in the field using a clinometer. However, today we will assume that it is a 2% slope (the middle of the soil mapping unit description of 1 to 3% slope). Using Table 16.2, determine the LS factor given the length of slope you measured combined with a 2% slope. Note that if your value falls between two listed values in the table, it is recommended that you use either the higher value for slope gradient and/or length of slope, so as to calculate a more conservative erosion estimation (i.e. higher erosion rate).

Table 16.2. Values for Topographic Factor (LS).

Slope Gradient	**Slope Length (ft)**						
(%)	50	100	150	200	300	400	600
2	0.16	0.20	0.23	0.25	0.28	0.30	0.34
4	0.30	0.40	0.47	0.53	0.62	0.70	0.82
6	0.49	0.67	0.82	0.95	1.17	1.35	1.65
8	0.70	0.99	1.21	1.41	1.72	1.98	2.43
10	0.97	1.37	1.68	1.94	2.37	2.74	3.36
12	1.28	1.80	2.21	2.55	3.13	3.61	4.42
14	1.62	2.30	2.81	3.25	3.98	4.59	5.62
16	2.01	2.84	3.48	4.01	4.92	5.68	6.95
18	2.43	3.43	4.21	4.86	5.95	6.87	8.41
20	2.88	4.08	5.00	5.77	7.07	8.16	10.00

Table adapted from Jones et al. (1988) with permission.

Next, determine the average annual soil loss in tons/acre, assuming that no conservation practices are being used, and that conventional tillage (straight up and down the hill) is being used:

$$A1 = R \times K \times LS$$

$$A1 =$$

To compare that erosion rate to the "tolerable rate", commonly known as T, navigate to the "Soil Properties and Qualities" tab within the "Soil Data Explorer" tab in the Web Soil Survey. Click on "T Factor", then click on "View Rating" to populate the map. Scroll down to the bottom of the page to see the rating in tons per acre per year.

How does the erosion rate under conventional tillage compare to the tolerable erosion rate?

The farmer wants to reduce the annual erosion rate from this slope, and needs your help to compare the impacts of potential conservation practices. One possibility is to switch to a conservation tillage practice in a wheat-on-wheat rotation that leaves at least 30% of the soil surface covered at the time of planting the next crop. Use Table 16.3 to determine the C factor for the wheat-on-wheat rotation if 30% of the soil surface is covered with residue at the time of planting the following crop under conservation tillage.

Table 16.3. Cover and management (C) values for combinations of tillage and residue cover after planting and crop sequence for wheat rotations.

	Tillage	Tillage	Tillage	Tillage	Tillage	Tillage	No Tillage	No Tillage	No Tillage	No Tillage	No Tillage	No Tillage
Residue	<5% (Fall)	<5% (Spring)	20%	30%	40%	50%	20%	30%	40%	50%	60%	70%
Rotation												
W-B		-		-	-	-	-	-	-	0.07	0.06	-
W-M		-	-	-	-	-	-	-	-	-	-	-
W-W	0.20	0.20	0.10	0.09	0.08	0.07	-	-	-	-	-	0.04
W-O	0.23	0.23	0.12	0.11	0.09	0.08	-	-	-		0.05	0.04
W-Co	0.16	0.16	0.13	0.11	0.10	0.08	-	-	-	0.06	0.05	-
Co-W	0.34	0.30	0.14	0.11	0.10	0.09	0.14	0.13	0.11	0.10	0.08	0.06
B-W	0.30	0.28	0.18	0.16	0.12	-	-	-	-	-	-	0.03

Table adapted from Jones et al. (1988) with permission.

† Sorghum (milo) may be substituted for corn; all C values are for wide row plantings.

‡ Crop abbreviations are as follows: Co, corn; B, soybeans; W, winter wheat; M, meadow (alfalfa, clover, grass, etc.); Fl, fallow.

Using the values from the previous RUSLE calculation (A1) above, incorporate the C factor to determine the resulting erosion rate after the implementation of conservation tillage.

$$\mathrm{A2} = \mathrm{R} \times \mathrm{K} \times \mathrm{LS} \times \mathrm{C}$$

$$\mathrm{A2} =$$

How does the erosion rate under conservation tillage compare to the tolerable erosion rate?

How does the erosion rate under conservation tillage compare to the erosion rate under conventional tillage?

The farmer wants to know what options are available that don't require the purchasing of new equipment (such as a planter or drill designed for planting through stubble). Determine the Pc factor for contour farming using Table 16.4 in the using the 2% slope for our field

Assuming conventional tillage will still be used (no C factor), calculate the erosion rate if just contour tillage was used, using the same R, K, and LS values from the previous RUSLE calculation above (A2).

$$\mathrm{A3} = \mathrm{R} \times \mathrm{K} \times \mathrm{LS} \times \mathrm{C} \times \mathrm{Pc}$$

$$\mathrm{A3} =$$

Table 16.4. Conservation practice (P) values for contour farming and contour strip

cropping.

	Contour Farming	Contour Farming	Contour Strip Cropping	Contour Strip Cropping	Contour Strip Cropping
Slope Gradient (%)	Max Slope Length (ft)	P Value	Strip Width (ft)	P Value, RGMM	P Value, RRGM
1 - 2	400	0.6	130	0.30	0.45
3 - 5	300	0.5	100	0.25	0.38
6 - 8	200	0.5	100	0.25	0.38
9 - 12	120	0.6	80	0.30	0.45
13 - 16	100	0.7	80	0.35	0.52
17 - 20	100	0.8	60	0.40	0.60

Table adapted from Jones et al. (1988) with permission. †Strip cropping uses a four-year rotation of row crop followed by one year of a small grain and two years of meadow (forages) for RGMM, or uses two years of row crops followed by one year of small grain and one year of meadow for RRGM. Meadow includes alfalfa, clover, grass, etc.

How does the erosion rate under contour tillage compare to the tolerable erosion rate?

How does the erosion rate under contour tillage compare to the erosion rate under conservation tillage alone?

Next we will test the impact of installing terraces on the landscape. Using Table 16.5, determine the Pt factor. When terraces are installed, contour tillage is usually used as well. Also, note that installing a terrace results in a shorter length of the slope (because the terrace stops water from continuing to run down slope), so this calculation is performed for each terrace individually. Also note that the net P factor is determined by multiplying the Pc and Pt values together, or writing the RUSLE as follows:

$$A4 = R \times K \times LS \times Pc \times Pt$$

Table 16.5. Conservation practice (P) values for terraces with underground outlets or waterways.

Terrace Interval	Underground Outlets	Waterways with percent grade of:		
(ft)		0.1-0.3	0.4-0.7	0.8
	Pt Values	Pt Values	Pt Values	Pt Values
<110	0.5	0.6	0.7	1.0
110-140	0.6	0.7	0.8	1.0
140-180	0.7	0.8	0.9	1.0
180-225	0.8	0.8	0.9	1.0
225-300	0.9	0.9	1.0	1.0
300+	1.0	1.0	1.0	1.0

Table adapted from Jones et al. (1988) with permission.

Assume that one terrace has a length of slope of 130 ft, has the same percent slope as above (2%), uses underground outlets to remove excess water, and that contour tillage practices are being used as described above, calculate the erosion rate for that single terrace.

$$A4 =$$

How does that erosion rate compare to the tolerable erosion rate?

How does that erosion rate compare to the likely geologic erosion rate of 1 ton/ac/yr?

Now, calculate the erosion rate if conservation tillage, contour tillage, and terraces were all used together for soil conservation as they were described above.

$$A5 = R \times K \times LS \times C \times Pc \times Pt$$

$$A5 =$$

How does that erosion rate compare to the tolerable erosion rate?

How does that erosion rate compare to the likely geologic erosion rate of 1 ton/ac/yr?

Considering your results, would installing terraces in addition to using contour tillage and conservation tillage make economic sense if the goal is to reduce erosion rates to near-geologic rates?

Describe how the USLE model could be used to make management decisions regarding tillage practices or terracing – specifically, how could a soil conservationist work with a farmer to reach a truly sustainable erosion rate in the easiest and most economically feasible way possible.

Activity 3: Preventing Wind Erosion

Early on, the estimation of wind erosion rates relied on the Wind Erosion Equation that was similar to RUSLE. It is as follows:

$$E = F \times I \times C \times K \times L \times V$$

Where E is the annual wind erosion rate in tons/ac, I is the soil erodibility factor, C is the climate factor, K is the soil-ridge-roughness factor, L is the width of field factor, and V is the vegetative cover factor. While this model does do a good job of visualizing the primary factors that control wind erosion, the accuracy of the prediction of wind erosion rates has greatly increased through computer based models, such as the Revised Wind Erosion Equation (RWEQ) and the Wind Erosion Prediction System (WEPS). However, as you can see, vegetative cover, surface roughness, and length of the field are all factors that can be manipulated by the farmer to reduce wind erosion.

Describe at least one way in which surface roughness can be managed to reduce wind speed, and thus wind erosion.

Describe how windbreaks and shelterbelts work, and how they must be oriented relative to the prevailing wind.

Strip-cropping is one conservation practice used to reduce water erosion. It is also used in reducing wind erosion. Describe strip cropping, and compare and contrast how the practice is utilized for reduction of wind erosion, and reduction of water erosion.

Assignment: Online Quiz

A quiz for this lab will be available online. Please access it as directed by your instructor.

References

Abbreviations

NCSS – National Cooperative Soil Survey
NRCS – Natural Resources Conservation Service
USDA – US Department of Agriculture
KSRE – Kansas State Research and Extension

References

2012rc. 2009. The Periodic Table of the Elements. https://commons.wikimedia.org/wiki/File:Periodic_table_large.svg (accessed 8 June 2017).

Allison, E., and C. Jones. 2005. Module 5 – Plant Water Relations. *In* Soil and Water Management, 12. Bozeman, MT: Montana State University. http://landresources.montana.edu/swm/documents/Final_SW_5_proof_11_18_05.pdf (accessed 7 June 2018).

Brady, N.C., and R.R. Weil. 2009. Elements of the Nature and Properties of Soils. 3rd Edition. Pearson, Upper Saddle River, N.J.

Brady, N.C., and R.R. Weil. 2017. The Nature and Properties of Soils, The, 15th Edition. 15th ed. Pearson, Upper Saddle River, N.J.

Cheng, X., R. Zhao, Q. Zhang, and H. Tjugito. 2014. Raindrop Impact on a Sandy Surface. American Physical Society. https://gfm.aps.org/meetings/dfd-2014/5407aca969702d0771ed0100 (accessed 6 July 2017).

CropWatch – Youth. 2013a. Soil Bulk Density Overview. https://www.youtube.com/watch?v=_ZVIIIlo5js (accessed 2 January 2020).

CropWatch – Youth. 2013b. Soil Bulk Density test. https://www.youtube.com/watch?v=bf9qTZQGNHs (accessed 2 January 2020).

CropWatch – Youth. 2013c. Soil pH Overview. https://www.youtube.com/watch?v=TSkjXGp8bAw (accessed 2 January 2020).

CropWatch – Youth. 2013d. Soil pH Test. https://www.youtube.com/watch?v=jFQeDef6bug (accessed 2 January 2020).

CropWatch – Youth. 2013e. Soil Respiration Overview. https://www.youtube.com/watch?v=be51efbavTo (accessed 2 January 2020).

CropWatch – Youth. 2013f. Soil Respiration Test. https://www.youtube.com/watch?v=RjnCf0weSUQ (accessed 2 January 2020).

Earle, S. 2015. Chapter 5 Weathering and Soil. *In* Physical Geology. Victoria, BC: BCcampus. https://opentextbc.ca/geology/part/chapter-5-weathering-and-soil/.

Fagerness, M.J. 2001. Fertilizing Kansas Lawns. Kansas State University Agricultural Experiment Station and Cooperative Extension Service, Manhattan. https://www.bookstore.ksre.ksu.edu/pubs/mf2324.pdf.

Hsieh, J.C., W.H. Gardner, and Washington State University. 1961. Water Movement in Soils. Washington State University. https://www.youtube.com/watch?v=WNYqkPaJd-I (accessed 7 June 2017).

Jones, A.J., D. Walters, W.G. Hance, E.C. Dickey, and J.R. Culver. 1988. Universal Soil Loss Equation: A Handbook for Nebraska Producers. Nebraska Cooperative Extension Service, Lincoln. http://digitalcommons.unl.edu/cgi/viewcontent.cgi?article=2620&context=extensionhist (accessed 5 June 2017).

King, L.D., H.J. Kleiss, and J.A. Thompson. 2003. SSC 200: Soil Science Laboratory Manual. North Carolina State University, Raleigh, NC, USA.

Kansas State University Department of Agronomy. 2020, January 13. Soil Analysis. https://www.agronomy.k-state.edu/services/soiltesting/farmer-services/soil-analysis/index.html (accessed 13 January 2020).

KSREVideos. 2010. Eye on Agriculture Today: Soil Texture By Feel. Kansas State Research and Extension. https://www.youtube.com/watch?v=IOyaBxj767s&t (accessed 26 December 2019).

KSREVideos. 2011a. Soil Profiling: Color. Manhattan, KS. https://www.youtube.com/watch?v=2cmJa8dQ5_I (accessed 26 January 2020).

KSREVideos. 2011b. Soil Profiling: Structure. Manhattan, KS. https://www.youtube.com/watch?v=UKT9RBIkeKc (accessed 26 December 2019).

KSREVideos. 2011c. Soil Profiling: The Proper Tools. https://www.youtube.com/watch?v=UtgTUXl4HMM (accessed 13 January 2020).Leikam, D.F., R.E. Lamond, and D.B. Mengel. 2003. Soil Test Interpretations and Fertilizer Recommendations. Kansas State University Agricultural Experiment Station and Cooperative Extension Service, Manhattan. https://www.bookstore.ksre.ksu.edu/pubs/MF2586.pdf.

Lindbo, D.L., D.A. Kozlowski, and C. Robinson. 2012. Know Soil, Know Life. Soil Science Society of Agronomy, Madison, WI.

McCauley, A., and C. Jones. 2005a. Module 3 – Managing for Soil Erosion. *In* Soil and Water Management, 12. Bozeman, MT: Montana State University. http://landresources.montana.edu/swm/documents/Final_proof_SW3.pdf.

McCauley, A., and C. Jones. 2005b. Module 4 – Water and Solute Transport in Soils. *In* Soil and Water Management, 12. Bozeman, MT: Montana State University. http://landresources.montana.edu/swm/documents/final_SW4_proof_11_18_05.pdf.

Montana NRCS. 2020. Soil Health Demonstration: Slake Test. https://www.youtube.com/watch?v=IBjeI_tCvAg (accessed 19 January 2021).

Moorberg, C. 2020. Agronomy Compost Facility. Manhattan, KS. https://www.youtube.com/watch?v=rGR1lbOaX9o&feature=youtu.be (accessed 14 January 2021).

NCSS. 1997. Official Series Description: Harney Series. https://soilseries.sc.egov.usda.gov/OSD_Docs/H/HARNEY.html (accessed 7 June 2017).

NCSS. 2012. Official Series Description – Colby Series. https://soilseries.sc.egov.usda.gov/OSD_Docs/C/COLBY.html (accessed 7 June 2017).

NCSS. 2014. Official Series Description: Kenoma Series. https://soilseries.sc.egov.usda.gov/OSD_Docs/K/KENOMA.html (accessed 7 June 2017).

NCSS. 2015. Official Series Description: Cherokee Series. https://soilseries.sc.egov.usda.gov/OSD_Docs/C/CHEROKEE.html (accessed 7 June 2017).

NCSS. 2016. Official Series Description: Tivoli Series. https://soilseries.sc.egov.usda.gov/OSD_Docs/T/TIVOLI.html (accessed 7 June 2017).

NCSS. 2017. Official Series Description: Ulysses Series. https://soilseries.sc.egov.usda.gov/OSD_Docs/U/ULYSSES.html (accessed 7 June 2017).

officemmdivide. 2011. Wind Erosion (english version). https://www.youtube.com/watch?v=PQmon7Rj6ns (accessed 17 June 2017).

O'Geen, T. 2021. SoilWeb Apps. https://casoilresource.lawr.ucdavis.edu/soilweb-apps/ (accessed 19 January 2021).

Presley, D., and S. Thien. 2008. Estimating Soil Texture by Feel. 4. Manhattan, KS: Kansas State University Agricultural Experiment Station and Cooperative Extension Service. https://www.bookstore.ksre.ksu.edu/pubs/MF2852.pdf.

Rogers, D. H., J. Aguilar, I. Kisekka, P. L. Barnes, and F. R. Lamm. 2014. Soil, Water, and Plant Relationships. 8. Manhattan, KS: Kansas State University Agricultural Experiment Station and Cooperative Extension Service. https://www.bookstore.ksre.ksu.edu/pubs/L904.pdf.

Soil Survey Staff. 2010. From the Surface Down: An Introduction to Soil Surveys for Agronomic Use. Second. Lincoln, NE: USDA NRCS. https://nrcspad.sc.egov.usda.gov/DistributionCenter/pdf.aspx?productID=449 (accessed 19 January 2021).

Soil Survey Staff. 2015. Illustrated Guide to Soil Taxonomy. Version 2.0. Lincoln, NE: US Department of Agriculture, Natural Resources Conservation Service, National Soil Survey Center. https://www.nrcs.usda.gov/wps/PA_NRCSConsumption/download?cid=stelprdb1247203&ext=pdf (accessed 19 January 2021).

Soil Survey Staff. 2021. Description of SSURGO Database. https://www.nrcs.usda.gov/wps/portal/nrcs/detail/soils/survey/geo/?cid=nrcs142p2_053627 (accessed 19 January 2021).

Sonon, L. S., U. K. Saha, and D. E. Kissel. 2017. Cation Exchange Capacity and Base Saturation. 4. Athens, GA: University of Georgia Extension. https://extension.uga.edu/publications/detail.html?number=C1040&title=Cation%20Exchange%20Capacity%20and%20Base%20Saturation.

Thompson, A., and K. W. Goyne. 2012. Introduction to the Sorption of Chemical Constituents in Soils. Nature Education Knowledge 4(4):7. https://www.nature.com/scitable/knowledge/library/introduction-to-the-sorption-of-chemical-constituents-94841002/.

Travel Links Directory. 2013. Rainfall Slow Motion HD Heavy Rain Drops Falling in Slow Mo Video View of Droplets Hitting Water. https://www.youtube.com/watch?v=tIJQkR-ofFo (accessed 7 June 2017).

USDA-NRCS. 1998. Soil Quality Resource Concerns: Soil Erosion. 2. Lincoln, NE: U.S. Department of Agriculture. https://nrcspad.sc.egov.usda.gov/DistributionCenter/pdf.aspx?productID=391.

USDA-NRCS. 2008a. Available Water Capacity. 2. Washington, D.C.: USDA NRCS. https://www.nrcs.usda.gov/wps/PA_NRCSConsumption/download?cid=nrcs142p2_051590&ext=pdf.

USDA-NRCS. 2008b. Bulk Density. 2. Washington, D.C.: USDA-NRCS. https://www.nrcs.usda.gov/wps/PA_NRCSConsumption/download?cid=nrcs142p2_051591&ext=pdf.

USDA-NRCS. 2009. Soil Respiration. 2. Washington, D.C.: USDA NRCS. https://www.nrcs.usda.gov/wps/PA_NRCSConsumption/download?cid=nrcs142p2_052801&ext=pdf.

USDA-NRCS. 2011. Soil pH. 2. Washington, D.C.: USDA NRCS. https://www.nrcs.usda.gov/wps/PA_NRCSConsumption/download?cid=nrcs142p2_052474&ext=pdf.

USDA-NRCS. 2014a. Soil pH. 6. Washington, D.C.: USDA NRCS. https://www.nrcs.usda.gov/Internet/FSE_DOCUMENTS/nrcs142p2_051574.pdf.

USDA-NRCS. 2014b. Soil Respiration. 9. Washington, D.C.: USDA NRCS. https://www.nrcs.usda.gov/Internet/FSE_DOCUMENTS/nrcs142p2_051573.pdf.

USDA-NRCS. 2016. Web Soil Survey. https://websoilsurvey.sc.egov.usda.gov/App/HomePage.htm (accessed 7 June 2017).

USDA-NRCS. 2017. Official Soil Series Descriptions (OSDs). https://www.nrcs.usda.gov/wps/portal/nrcs/detail/soils/home/?cid=nrcs142p2_053587 (accessed 16 June 2017).

USDA-NRCS. 2019. Soil Bulk Density/Moisture/Aeration. 11. Washington, D.C.: USDA-NRCS. https://www.nrcs.usda.gov/Internet/FSE_DOCUMENTS/nrcs142p2_050936.pdf.

USDA-NRCS. 2020, January 9. Soil Formation and Classification. https://www.nrcs.usda.gov/wps/portal/nrcs/detail/soils/edu/?cid=nrcs142p2_054278 (accessed 19 January 2020).

US EPA, O. 2015, August 19. Types of Composting and Understanding the Process. https://www.epa.gov/sustainable-management-food/types-composting-and-understanding-process (accessed 19 November 2019).

Upham, W., and C.W. Marr. 2011. Fertilizing Gardens in Kansas. Kansas State University Agricultural Experiment Station and Cooperative Extension Service, Manhattan. https://www.bookstore.ksre.ksu.edu/pubs/MF2320.pdf.

Whitney, D., and R. Lamond. 1993. Liming Acid Soils. 4. Manhattan, KS: Kansas State University Agricultural Experiment Station and Cooperative Extension Service. https://www.bookstore.ksre.ksu.edu/pubs/MF1065.pdf.

Woods End Laboratory. 2009. Solvita Compost Test: How to Perform it. https://www.youtube.com/watch?v=nD8O_TRN6bY&feature=emb_logo (accessed 14 January 2021).

Woods End Laboratory. 2019. Solvita Soil CO2 Respiration Updated Protocol Improvement. https://www.youtube.com/watch?v=7iZgzNZascE (accessed 19 January 2021).

Conversion Factors and Formulas

Conversion Factors

acre (ac)	=	**0.4 hectare(ha)**
acre-inch (ac-in)	=	27,154 gallon (gal)
cubic feet (ft3)	=	7.48 gallon (gal)
cubic centimeter (cm3)	=	0.000001 cubic meter (m3)
cubic feet (ft3)	=	0.028 cubic meter (m3)
cubic inch (in3)	=	16.39 cubic centimeter (cm3)
cubic yard (yd3)	=	0.76 cubic meter (m3)
feet (ft)	=	0.3048 meter (m)
gallon (gal)	=	3.79 liter (L)
gallon/acre (gal/ac)	=	9.35 liter/hectare (L/ha)
hectare (ha)	=	10,000 square meter (m2)
inch (in)	=	2.54 centimeter (cm)
mile (mi)	=	1.61 kilometer (km)
ounce – weight basis (oz)	=	28.35 gram (g)
ounce – fluid basis (oz)	=	30 milliliters (mL)
parts per million (mg/kg)	=	0.002 pound/ton (lb/t)
parts per million (mg/L)	=	0.00835 pound/1000 gallon (lb/1000gal)
percent (%)	=	10,000 parts per million (ppm)
pound (lb)	=	0.454 kilogram (kg)
pound/1000 gallon (lb/1000gal)	=	27.154 pound/acre-inch (lb/ac-in)
pound/acre (lb/ac)	=	1.12 kilogram/hectare (kg/ha)
pound/cubic foot (lb/ft3)	=	16.02 kilogram/cubic meter (kg/m3)
pound/gallon (lb/gal)	=	119,826 parts per million (mg/L)
square feet (ft2)	=	0.093 square meter (m2)
square mile (mi2)	=	2.59 square kilometer (km2)
square mile (mi2)	=	640 acre (ac)
ton (t)	=	2,000 pound (lb)
ton (t)	=	0.907 metric ton (Mg)
ton, metric (Mg)	=	2,205 pound (lb)
ton, metric (Mg)	=	1,000 kilogram (kg)
yard (yd)	=	0.9144 meter (m)

Formulas

Area of a square	**=**	**length x width**
Area of a circle	=	Pi x r2
Volume of a cube	=	length x width x height
Volume of a cylinder	=	h x Pi x r2

Periodic Table of the Elements

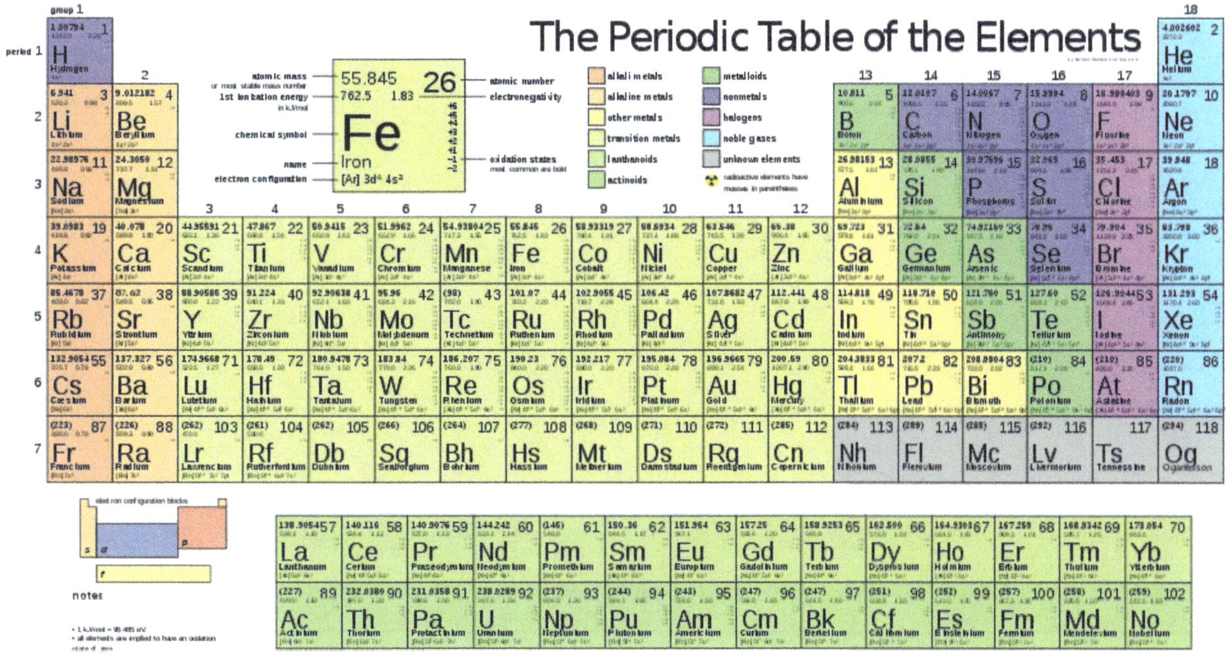

Figure courtesy of 2012rc (2009)

Made in the USA
Monee, IL
02 November 2026

67374724R00096